Do * it * yourself
Astronomy
* * *

Do * it * yourself
Astronomy

✳ ✳ ✳ ✳ ✳
✳ ✳ ✳
✳

Sydney G. Brewer

Edinburgh University Press

© Sydney G. Brewer 1988

Edinburgh University Press
22 George Square, Edinburgh

Set in Linotron Times Roman by
Pindar (Scotland) Ltd, Edinburgh and
printed in Great Britain by
Redwood Burn Limited
Trowbridge, Wilts.

British Library Cataloguing
  in Publication Data
Brewer, Sydney G.
DIY astronomy
1. Astronomy – Amateurs' manuals
I. Title
522

ISBN 0 85224 573 4

Contents

✳ ✳ ✳

Preface
✳ ✳ ✳

This book is based on some original experiments carried out in and around Edinburgh, with the general theme of measurement outwards from ourselves to the stars and including some historical and other material. The topics are traditional, but the use of very simple equipment gives considerable scope for innovation. Some mathematics has been used to derive formulae and to explain methods; it is mostly algebra, trigonometry and a little geometry but it stops short of serious calculus.

Some references to past and future events are enlivened with planetary positions found with an Acorn Electron computer using the excellent program in BASIC by Dr P. Duffett-Smith and published in book form as *Astronomy with your Personal Computer* by the Cambridge University Press.

My thanks are due to Dr D. W. Dewhirst of the Cambridge Observatories for his candid and helpful comments at an early stage in this project, to my former colleague John C. Naiff who read the final script, and to the ever-helpful staff of the National Library of Scotland.

The Edinburgh University Press gave support to the idea of this book: my sincere thanks go to all involved in the production.

Edinburgh 1988

Introduction

✳ ✳ ✳

In the gathering dusk of a March afternoon I turned off the main road near Dunbar in the east of Scotland and made my way through the quiet village of Spott which I knew to be roughly east of my home in Edinburgh. Taking advantage of an improvement in the weather, I was hoping to fit in some observations of a special pair of stars in the constellation Orion before the rapidly lengthening days made it impossible. In the last of the daylight I fitted together my simple equipment in the corner of a field and then had time to look around. It was indeed a scene to remember: the fading glow of sunset filled the western horizon and a three-day-old moon hung above it, while in the darker sky to the south the stars of Orion flashed in the crisp air. Then I saw some distance away an object by the roadside neatly fenced off and with a sign saying THE WITCHES' STONE. Somewhere in the distance an owl hooted. As I made my preparations a lone figure walked slowly up the lane – my arrival had evidently been noted.

'Just some astronomy' I ventured. The visitor gave a glance at the large tripod from which hung down a line into a bucket of water.

'Ah. I did wonder.' After a brief conversation he made off past the Witches' Stone towards a light in the distance and for the following hour or so I was left to complete my observations.

Had I been pressed for some technical explanation I would have said that I was trying to measure the radius of the earth – not perhaps a very convincing story. And why do this knowing perfectly well that the figure is 6400 kilometres or 4000 miles, very nearly? One might as well ask why people climb the rock faces of mountains when there is a perfectly easy grass slope to the summit on the other side, or why they are drawn to any outdoor pursuit with the elements of challenge and a direct contact with nature.

1

They do it because it is different, probably quite difficult, and a refreshing change from regular routine.

In astronomy nowadays it is rare for the amateur to be able to compete on equal terms with the professional in obtaining significant new results. But although the amateur may have limited means, he can with luck find in a project something original or unusual and in so doing will have his reward. Also, he is free to choose his own field of interest and to pursue it in his own style.

Long before that incident at Spott, I had the idea of setting aside nearly all advanced equipment including telescopes, and simply going out under the stars to rediscover from the most basic measurements as much astronomy as possible. This was not in any way to decry modern progress, nor was it a wish to relive the past, though it does bring one into touch with the history of the subject. It just seemed that a simple approach and a new look could yield interesting results. And for the measurements in mind telescopes would not often give a great advantage. One indispensable modern item was a watch or clock reading down to seconds and backed up by professionally produced radio time signals; also important was a pocket calculator with trigonometric functions. For lengths I used a steel foot ruler also calibrated in millimetres, and from it were derived all later distances by a process which I hope the reader will eventually judge to have been fairly honest. To cheat in a project of this kind removes much of the enjoyment, though of course it is interesting to verify results after they have been found and after devising the necessary experiments.

This book is an account of those experiments linking the foot ruler with the distances to the stars. Although practical descriptions are included, some readers will probably be content to remain comfortably in their armchairs and imagine it all taking place, while a few experimenters will, after thinking things out, prefer their own practical ideas, which is all to the good – the possibilities are almost endless.

Although there are various references to the latitude of Edinburgh which is 56° N the methods are readily adapted to other latitudes and a few diagrams have been drawn for 40° N as well as for 56° N. Admittedly we are rather far north up here and the

Introduction

distance from the north pole to the 18th green at St Andrews is actually less than that from the south pole to the dreaded Cape Horn, the southernmost point of any of the major land masses excepting Antarctica itself.

Metric units are used as a rule, but some expressions in general speech are slow to change and as this is not a textbook it seems unnecessary to insist on absolute uniformity. To me the 100 inch telescope at Mount Wilson will always be just that, and a pleasant walk to 3000 feet is somehow diminished when written as 914 metres!

Some of the more mathematical details have been placed at the ends of the chapters, together with a few practice exercises with answers. There are also two mathematical appendices which should not be beyond senior pupils at school, now that the scalar product of two vectors is commonly taught.

The book begins with a general astronomical introduction and two experiments to find the length of the sidereal day and the observer's latitude, both of which are important later. Then after considering Kepler's laws of planetary motion we embark on a first journey through the solar system and out as far as Saturn. Then we realise there is a problem about absolute distances which occupies two chapters. We then return to earth and catch up with some measurements both of the earth and of the moon; with these the investigation of the solar system is brought to a more satisfactory state. There follows a look at the nearer of the stars surrounding us and an attempt to find their distances. Finally, by one of the simplest yet strangest of experiments we begin to realise, even without seeing it all, the enormous extent of the whole universe.

I hope that after stepping aside from modern technology and considering afresh our privileged and yet lonely situation in its full perspective, the reader will feel support for those who seek to conserve and help the great variety of life on this planet; this is our urgent task which goes beyond all national barriers.

1

The sidereal day and the celestial sphere
✳ ✳ ✳

Figure 1.1   The earth to a scale of 1 millimetre = 1 kilometre

The sidereal day

In these days of world-wide tele-
vision coverage and rapid air travel there can be an impression
that the earth is not so vast after all. Let me anticipate one of the
results of this project and quote the radius of the earth as I measured it in 1984: it was 6360 kilometres which is quite close to the
correct value. Imagine a model perfect in every detail constructed
on a scale of one to a million, with 1 millimetre representing 1
kilometre. The earth would be a globe with diameter 12·72 metres,
roughly 40 feet and higher than many houses. Almost the entire
atmosphere would be within a few inches of the surface; indeed
the weather-producing clouds would mostly be within half an inch
of the surface. Peering at this great globe from a distance of a foot

4

one would see some evidence of the presence of man: the largest cities a couple of inches across, at night glowing with a soft light, by day revealing arterial routes like spider threads. Far out to sea would be the tracks of supertankers, themselves less than a millimetre long.

So the earth does in fact contain an enormous amount of material and as a result it spins in space about its axis with very great regularity, being hardly affected by weather movements within its atmosphere or tidal effects in its oceans or shifts within its crust. Even the activities of some thousands of millions of human beings and their machines in constant motion over its surface have no noticeable effect. And this spin takes place against the background of the stars which are so far away that the constellations appear unchanged over our lifetime.

We see this steady rotation relative to the stars as an equal and opposite apparent motion of the celestial background about an axis parallel to the earth's axis, the time of one revolution being known as one sidereal day. The sidereal day is important for our purposes because it is easy to measure and practically constant. It is a little shorter than the solar day which controls our lives and which is the time taken by the earth to revolve once relative to the sun.

To see the difference between the two kinds of day, place a chair in the middle of the room and stand facing it and also facing a particular wall. Walk right round the chair once, spinning round in the same direction a number of times as you go. Count the times after the start of all this that you (the earth) face the chair (the sun) during the complete circuit (one year). Also have someone count the times after the start that you (the earth) face the wall (a distant constellation) during the circuit (one year). If you concentrate hard you may be able to do both counts simultaneously yourself! The answer for the stars you will find is one more than for the sun, because in the process you have encircled the sun once. Likewise a point on the real earth is brought face to face with the sun 365 times in a year, with a few hours left over, but face to face with a particular star 366 times in a year, with a few hours left over. In order to make that extra revolution in a year the stars rise and set about four minutes earlier in every sun-controlled day.

The sidereal day we are trying to find is therefore about 23 hours 56 minutes of mean solar time. Because of the steady rate of rotation, if we time to the nearest second the total of many revolutions then by division we have a very accurate value for the time of one revolution.

Any bright star not too near the celestial pole and disappearing or reappearing from behind any convenient fixed object can be used to obtain a good result for the length of the sidereal day. From a window in my house I have a clear view of the bright northern star Capella as it reappears from behind a neighbour's chimney stack. I glued a small square of black paper on the window, with a central 6 mm viewing hole, and persuaded the family to leave it well alone for a few months. Then as the weather allowed it was easy to time the star's reappearances to the nearest second; by keeping a lookout to the right of the paper square there was plenty of warning of the star's approach. It would have been a little easier to time a disappearance, but from my viewpoint this was impossible.

Here are some results which I obtained in this way in 1984. They are corrected for watch error thanks to those who produce the high-quality radio time signals which we take so much for granted. Also one must remember that over a long period the start or finish of Summer Time may have to be allowed for.

(1) 43 revolutions in $3\,705\,056^s$ giving a sidereal day of $86\,164^s \cdot 093$

(2) 72 revolutions in $6\,203\,816^s$ giving a sidereal day of $86\,164^s \cdot 111$

Either of these rounded to the nearest tenth of a second is $86\,164^s \cdot 1$ or $23^h\ 56^m\ 04^s \cdot 1$, the value we shall adopt.

This experiment, which on a winter's night has the considerable advantage of being staged indoors, is well worth repeating to see at first hand the silent precision with which the star appears or disappears, always at the same spot and always on time. It also serves as an introduction to the technique of observing used later on. It might be thought that a smaller hole would lead to sharper

The sidereal day and the celestial sphere

timing, but it is better to have plenty of starlight reaching the eye; indeed in city surroundings every bit is needed, so the hole should be about as big as the eye pupil. If a disappearance is being observed (and subsequently we shall always arrange this), keep the star in the middle of the little field formed by the out-of-focus outline of the hole, avoid staring directly at the star and try to judge the moment when the light is fading fastest – some practice helps. It is a mistake to wait until the last trace of light is gone; on a hazy night it will vanish earlier while with a bright star on a clear dark night the instant may be delayed by diffraction at the edge of the obstruction. So the moment of rapid change is best. Do not expect a timing better than plus or minus 1 second for this deliberately simple experiment. The real accuracy stems from taking measurements extending over many sidereal days, as stated earlier.

The Celestial Sphere

Standing under the stars on a clear night and well away from city lights it is easy to imagine that one is at the centre of a vast black sphere which, as we have seen, rotates once every sidereal day and to which the stars, moon and planets seem attached. Just as places on earth are fixed by their latitude and longitude so heavenly bodies are located by a similar coordinate system which rotates with this celestial sphere. Imagine the earth to be small and transparent with a light source at its centre (Figure 1.2), then the circles of latitude would be projected outwards onto the celestial sphere where they are called circles of declination; our equator and poles would project into the celestial equator (declination 0°) and north and south celestial poles (declinations 90° and −90°).

The north celestial pole (n) is overhead when seen from our north pole (N); it is on the north horizon when seen from our equator, and at any other place in the northern hemisphere its altitude angle above the north horizon equals the observer's latitude, as any sailor knows. There happens to be a fairly bright star

7

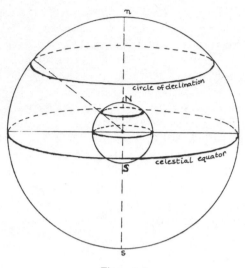

Figure 1.2

near the north celestial pole; it is for that reason called Polaris and is useful in estimating latitude. But there is no conspicuous star near the south celestial pole for the benefit of southern observers, a long-suffering race who also have to put up with astronomy books often written from a northern point of view.

The second coordinate in the heavens is called right ascension (RA) and corresponds to longitude on earth. Just as longitude is reckoned from Greenwich, England, so right ascension is reckoned from the First Point in Aries where the sun's annual path through the stars (the ecliptic) crosses the equator from south to north. Right ascension can be given in angular measure increasing eastwards from 0° to 360°, but is often expressed in time measure from 0^h to 24^h. If two stars differ in right ascension by three hours then, no matter what their declinations, they will cross the observer's meridian plane (through himself and the poles of the earth) three hours apart; but notice the hours will be sidereal hours, each one twenty-fourth of a sidereal day and therefore a little shorter than three hours of familiar mean solar time. As there are

The sidereal day and the celestial sphere

$24 \times 60 \times 60 = 86\,400$ seconds in a mean solar day, the interval between the two stars would be $\dfrac{86\,164 \cdot 1}{86\,400} \times 3\,\mathrm{h} = 2^{\mathrm{h}}\,59^{\mathrm{m}}\,30^{\mathrm{s}} \cdot 5$ of mean solar time.

Viewing the solar system from the outside one would notice that the planets evidently share with the sun a common origin, for though widely spread in space they are never far from an invariable flat plane which passes through the sun. Seen from the earth the other planets are therefore confined to a flat belt of sky called the zodiac which extends some 8° on each side of the ecliptic.

Figure 1.3(a) shows the spinning earth in relation to the plane of its orbit round the sun in the position corresponding to northern summer. The sun must be imagined far out to the left; to scale it would be a 3 metre diameter globe at a distance of 350 metres. The plane of the equator is inclined at $23\frac{1}{2}°$ to the plane of the orbit.

Figure 1.3(b) shows the celestial sphere with the earth reduced to a small blob at the centre. The direction of spin is here reversed because it is relative to the central earth; the celestial equator (parallel to the earth's equator) cuts the ecliptic (coincident with the plane of the orbit) again at $23\frac{1}{2}°$. At intervals round the ecliptic are twelve reference letters. Seen from the central earth, the sun is still out to the left and appears at D on the celestial sphere.

The next diagram (Figure 1.4) is divided into two for convenience and shows on a cylindrical projection the equatorial region of the celestial sphere together with the ecliptic. The letters on the ecliptic correspond to those on the previous diagram, A being the First Point in Aries. The dominant feature in this part of the heavens is the constellation Orion, very conspicuous in winter, and straddling the equator at around $5^{\mathrm{h}}\,30^{\mathrm{m}}$ right ascension.

The following description is both for an observer at latitude 56° N and for another at 40° N (with references in brackets).

The celestial equator intersects the horizon at all times in the east and west points, and at an angle of $90° - 56° = 34°$ $(90° - 40° = 50°)$.

In April the sun has reached B and just after sunset the western horizon is the broken line PQ (P′Q′). Turn the diagram until PQ

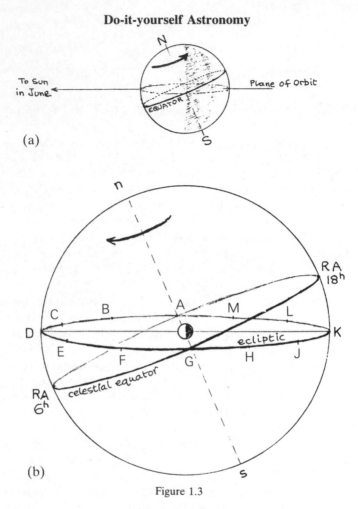

(a)

(b)

Figure 1.3

(P'Q') is horizontal; Orion is setting towards the sunset and in a few weeks will be lost as the days rapidly lengthen. By mid August the sun has reached F and at the first hint of dawn the eastern horizon lies along RS (R'S'); turn the diagram again and see how Orion has reappeared in the south-east though the sun has still to rise. For anyone chancing to be awake early, this reappearance of Orion is one of the great sights in the annual round: there is

The sidereal day and the celestial sphere

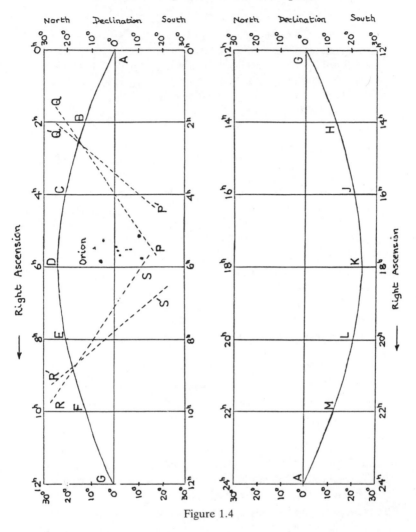

Figure 1.4

probably a nip in the air and a hint of the coming of autumn and winter. See how steeply the ecliptic slopes down to the horizon giving the best chance of seeing Venus or Mercury if they happen to be morning stars. Also in the same part of the zodiac there may

11

be the thin crescent of the waning moon, with the rest of the lunar disc dimly lit by earthshine reflected from eastern countries where the new day is already well advanced.

The sun, moon and planets move slowly round the zodiac always following accurately or approximately the direction of the ecliptic; for future reference it is useful to estimate the *zodiacal angle of climb (or descent)*. At A and G it is 23·5°, at B, F, H and M it is 20°, at C, E, J and L it is 12° and at D and K it is zero. Rough estimates are sufficient as only the cosines of these angles will be needed and they are all between 0·9 and 1·0 and mostly bunched close to 1·0.

We come now to some other preliminaries mainly to do with angles. Angles in degrees are subdivided into arcminutes (') and arcseconds (") formerly known as minutes of arc and seconds of arc and which are related thus: $1° = 60'$ and $1' = 60''$.

There is a useful rule sometimes called the 'one in sixty rule': an object with width 1 unit when viewed from a distance of sixty units presents or subtends at the eye an angle of very nearly 1°, and so on in proportion. It happens that when a scale marked in centimetres is held at arm's length it is about 60 cm from the eye and may be used to estimate angles in the sky up to a few degrees, for example the separation of the Pointer stars in Ursa Major is found to be about 5°. Again, if a ruler clearly divided into millimetres is set up at a distance from the eye of $60 \times 60 = 3600$ mm which is about 12 feet, the divisions will subtend angles of 1' (1 arcminute).

Most of us could not claim to see any detail of shape as small as this, such as the thin bright crescent of Venus which is so close to the limit of resolution when it is nearest to the earth. I have tried scribing thin crescents with diameters 2 mm, 4 mm, 6 mm side by side on a piece of smoked glass and then viewing them from a distance of 24 feet with a bright sky beyond so they shone like planets with diameters 1', 2', 3'. I could see the shape of the biggest, the middle one was doubtful and the small one impossible. Of course we can detect thin threads such as spiders' webs with angular widths much less than 1', but this is not the same as being certain of shape in a small disc or crescent.

12

The sidereal day and the celestial sphere

As an example of a moderately small angle, the middle star Mizar in the tail of Ursa Major has a fainter neighbour Alcor at a distance of 11′. This is a good preliminary test for anyone thinking of attempting some of the experiments in this book – try it on a dark night when Ursa Major is well clear of the horizon. The two stars are not really associated, they just happen to be nearly in line with us. But Mizar itself is a genuine double star which shows up well in quite a small telescope; its two components are separated by 14″ which will slowly change over the years as they revolve about each other.

Here is a summary concerning the speed of rotation of the celestial sphere. It is accurately 360° in 24^h of sidereal time; it is approximately 360° in 24^h of mean solar time with an error of about 4^m. This rate is 15° in 1^h; 1° in 4^m; 1′ in 4^s. For small intervals of angle it is sufficient to apply this rate using mean time as given by radio signals.

The diameters of the sun and moon are by a coincidence both about 30′; they are never very far from the celestial equator and so each appears to move through its own diameter in 2 minutes.

Stars close to the celestial equator move at the rate of 1′ in 4^s but others well away from the equator such as the Ursa Major group move more slowly, just as when we spin a terrestrial globe Greenland moves more slowly than Africa. To be exact, the speed at declination δ is reduced by multiplying by cos δ; thus the more northerly Pointer star Dubhe at declination 62° moves at a rate of $1 \times \cos 62° = 0.47$ arcminute in 4 seconds of time, as seen by the observer at the centre of the celestial sphere.

Finally, in Figure 1.5 are two diagrams, one for latitude 56° N and the other for latitude 40° N (and for which subsequent figures are in brackets). The observer is O on a flat plain; N, E, S, W are the north, east, south and west points of his horizon, Z is his zenith and n is the north celestial pole at altitude 56° (40°) above the north horizon, this being also the latitude of the place.

B represents a star near the celestial equator such as one of the line of three forming the belt of Orion, or it might be the sun on 21 March or on 21 September; such an object rises in the east, crosses the observer's south point at an altitude 90° − 56° = 34°

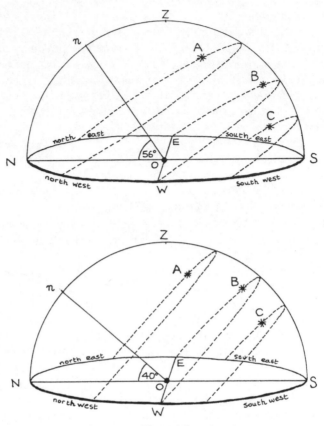

Figure 1.5

$(90° - 40° = 50°)$ and later sets in the west having been visible for 12 hours.

Again, A represents a star with northern declination 19° such as the bright red star Arcturus, or it might be the sun on 16 May or on 28 July; such an object rises to the north of east, crosses the south point at an altitude $34° + 19° = 53°$ $(50° + 19° = 69°)$ and later sets to the north of west having been visible for more than 12 hours.

Lastly, C represents a star with southern declination 17° (also

The sidereal day and the celestial sphere

written −17°) such as Sirius, the brightest fixed star in the sky, or it might be the sun on 2 February or on 10 November; such an object rises to the south of east, crosses the south point at an altitude 34°−17°=17° (50°−17°=33°) and later sets to the south of west having been visible for less than 12 hours.

Notes on Chapter 1

Note 1.1. Short distances on the celestial or terrestrial spheres may be found with sufficient accuracy using Pythagoras' result. With terrestrial diagrams we look in to the centre, as with a map; with celestial diagrams we look out from the centre. On the earth's surface 1 nautical mile subtends or presents at the centre of the earth an angle of 1 arcminute. To convert time to angle use $1^m=15'$.

Example 1. Find the separation in arcminutes between the stars Castor and Pollux.

      Castor: Right ascension $7^h 33^m$ Declination 31° 55′ N
      Pollux: RA $7^h 44^m$ Dec.   28° 04′ N.
      Change of dec.=3° 51′=231′=CZ (see Figure 1.6(a))
      Change of RA=11^m=11×15′=165′ (converting to angle)
      Multiply by the cosine of mean declination to allow for the crowding of RA lines when away from the equator:
      PZ=165 cos 30° arcminutes at the centre of the sphere.
      PZ=143′
      Therefore CP=$\sqrt{(231^2+143^2)}$=272 arcminutes.

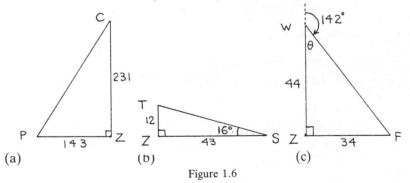

Figure 1.6

15

Example 2. The coordinates of Saturn on two occasions in 1962 were:

    1 August RA 20^h 43^m Dec. 18° 54′ S

    11 August RA 20^h 40^m Dec. 19° 06′ S

Find the movement in arcminutes in the ten days, and the zodiacal climb/descent angle.

    Change of dec.=12′=ZT (see Figure 1.6(b))

    Change of RA=3^m; SZ=3×15×cos 19°=43′ (done as in Example 1)

    Therefore ST=$\sqrt{(12^2+43^2)}$=45 arcminutes

    Also the climb/descent angle is arctan (12/43)=16°.

Example 3. Find the length in nautical miles of the sea passage between Fraserburgh,   Longitude 2° 00′ W   Latitude 57° 42′ N

    and Wick,   Longitude 3° 05′ W   Latitude 58° 26′ N.

    Change of latitude=44′=WZ (see Figure 1.6(c))

    Change of longitude=65′

Multiply by the cosine of mean latitude to allow for the crowding of longitude lines when away from the equator:

    ZF=65 cos 58°=34′ subtended at the centre of the earth.

    Therefore WF=$\sqrt{(44^2+34^2)}$=56′; as a distance this is 56 nautical miles.

    Also angle θ=arctan (34/44)=38° so the true course to steer from Wick to Fraserburgh is 142°.

Example 4. Find the separation in arcminutes between the Southern Cross stars, Alpha Crucis and Beta Crucis.

    Alpha Crucis: RA 12^h 25^m Dec. 62° 59′ S

    Beta Crucis: RA 12^h 46^m Dec. 59° 35′ S.

                                (Answer: 254′)

Example 5. The coordinates of Jupiter on two occasions in 1962 were:

    11 Sept. RA 22^h 35^m Dec. 10° 30′ S

    21 Sept. RA 22^h 30^m Dec. 10° 54′ S.

Find the movement in arcminutes in the ten days, and the zodiacal climb/descent angle.

                        (Answers: 78′; 18°)

The sidereal day and the celestial sphere

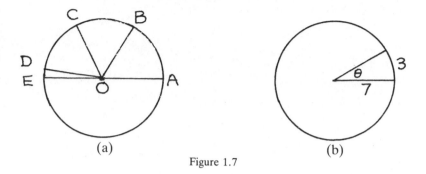

Figure 1.7

Example 6. Find the bearing and distance of Oslo from Kristian-
sand given the coordinates

    Oslo   Long. 10° 40′ E   Lat. 59° 55′ N,
    Kristiansand Long. 8° 00′ E   Lat. 58° 08′ N.
                           (Answers: 135 nautical miles; 037°)

Note 1.2. *Angles expressed in radians.* In Figure 1.7(a) the lengths
of the circular arcs AB, AC, AD are respectively one, two and
three times the length of the radius of the circle. The central angles
AOB, AOC, AOD then measure 1, 2 and 3 radians.

    Because the semicircular arc ABCDE has a length π times
the length of the radius, the central angle EOA measures π radians.
Therefore π radians=180 degrees, the basis of all conversions.
Thus 1 radian= $\dfrac{180}{\pi}$ degrees=57·3° nearly, and 1 degree = $\dfrac{\pi}{180}$
radians.

    In Figure 1.7(b),

$$\text{angle } \theta = \frac{3}{7}\text{ radians} = \frac{3}{7} \times \frac{180}{\pi}\text{ degrees}=24 \cdot 6°.$$

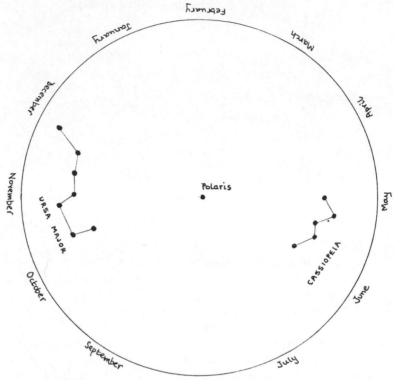

Figure 2.1   Polaris and the guide constellations. Turn the page until the current month is at the bottom. The figure then shows the appearance of the guide constellations Ursa Major and Cassiopeia on either side of the north celestial pole about four hours before local midnight

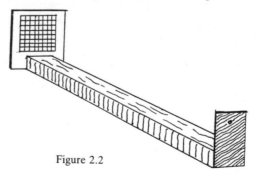

Figure 2.2

18

2

Polaris and the finding of latitude
* * *

Figure 2.1 shows some of the brighter stars near the north celestial pole; they are Polaris and the two constellations Cassiopeia and Ursa Major. By turning the diagram until the current month is at the bottom we have the appearance of this part of the sky for a northern observer about four hours before local midnight.

We know that latitude is given by the angle of altitude of the north celestial pole and that the position of Polaris is a rough guide, but without a star map how far is Polaris from the pole, and in what direction? The first experiment to be described gives approximate answers to these questions by tracking Polaris over a period of some hours.

On a sheet of clear perspex is scribed a grid of 1 cm squares, eight each way is more than adequate. The sheet is fixed perpendicularly at the end of a solid length of wood with a 4 mm viewing hole at the other end, as in Figure 2.2. The grid is made visible by rubbing a touch of white paint into the lines and cleaning off the surplus. The length of the 'instrument' is 120 cm which ensures that the squares when viewed through the hole appear with side lengths of half a degree or 30′ against the sky, by the one in sixty rule.

The device is very firmly mounted with one set of lines horizontal and Polaris in view through the hole and somewhere near the middle of the grid. A red torch shining on the grid makes a good contrast with the star and a moonless night is preferable. If dew on the perspex is troublesome a warm dry cloth may be applied gently but, of course, without disturbing the position of anything. At intervals of about half an hour the star's coordinates within a particular square are estimated to the nearest tenth of a square each way. There is no need to record anything but the tenths, the

19

Movement of Polaris   1983 December 12

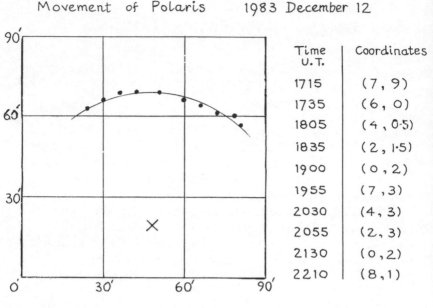

Time U.T.	Coordinates
1715	(7, 9)
1735	(6, 0)
1805	(4, 0·5)
1835	(2, 1·5)
1900	(0, 2)
1955	(7, 3)
2030	(4, 3)
2055	(2, 3)
2130	(0, 2)
2210	(8, 1)

Figure 2.3

change from one square to another is obvious when the numbers are plotted on graph paper. For greatest accuracy the eye should be placed so the star is in the middle of the out-of-focus outline of the hole. Figure 2.3 shows some results which I obtained in this way on 12 December 1983, together with the best-fitting circular arc as judged by eye. The times are given in Greenwich Mean Time which is also known as Universal Time (UT). Of course a determined observer on a clear winter's night could collect enough results for an arc extending over at least a semi-circle.

The experiment, rough as it was, produced two main conclusions:

(i) Polaris has its greatest altitude (called upper culmination) when the W-shaped constellation Cassiopeia has just passed through the meridian, near the zenith in Edinburgh. In other words, Polaris is on the Cassiopeia side of the north celestial pole.

(ii) The angular distance of Polaris from the pole, which is

20

the radius of the above circular arc, is about 49', rather less than twice the diameter of the full moon.

With these results as a guide there followed a second more accurate experiment to find the local latitude needed for subsequent calculations. Figure 2.4 shows the arrangement: a very rigid triangular sighting frame is bolted to a strong wooden upright and held to a convenient drainpipe by elastic ropes. The triangle can therefore move like the sail of a dinghy, but once Polaris is sighted near its culmination in the direction CA the frame is firmly fixed in the plane of the meridian, that is with C south of B, and so it remains. The triangle should really be in a vertical plane but a few millimetres out is not important.

In the figure, A is a horizontal steel needle, diameter 1·5 mm, from which hangs a vertical fishing line carrying a weight submerged in a bucket of water to stabilise it. On the way down, the line passes very close to the fixed scale; when viewed squarely this defines vertex B. On the short pivoted arm at C is a horizontal slit 2 mm wide; its position when read on the nearby scale defines vertex C. First, Polaris is sighted when at its greatest altitude at upper culmination, the little arm being moved until the slit is in line with needle and star, cutting off the light. There is plenty of time to make several independent checks and to find accurately the middle of the dark zone because the altitude hardly changes in half an hour around the maximum. It is all done when Cassiopeia is high and just past the meridian, remembering the result of the first experiment. Later the least altitude is found in the same way; if the weather allows this can be done after twelve hours by just moving the pivoted arm. In late January in Edinburgh there is enough darkness for everything to be completed in a single clear night, at about $18^h 15^m$ in the evening followed by $06^h 15^m$ the next morning. The average or mean of the two altitudes is the altitude of the celestial pole and therefore the observer's latitude. There will be a small error due to the refraction of the starlight as it enters the atmosphere, but without knowing the air's optical properties no allowance can be made.

Assisted by a calm clear night on 24–25 January 1984 and after measuring carefully the sides BC, CA, AB of the triangle for both

Do-it-yourself Astronomy

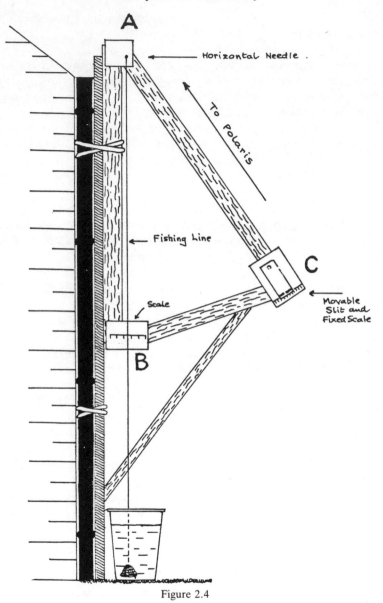

Figure 2.4

22

Polaris and the finding of latitude

upper and lower culminations, the greatest and least altitudes of Polaris came to 56° 47·4′ and 55° 11·9′ which have a mean value close to 56° 00′, the apparent latitude. This is actually in error by 2′ 30″ of which 40″ is due to refraction. Some further details are given at the end of this chapter. This idea of sighting with a vertical line is far from new; a similar device called a parallactic ladder was probably in use in the time of the great astronomer Ptolemy, around A.D. 140.

The distance of Polaris from the pole can now be found more accurately, for it is half the difference between the apparent altitudes at upper and lower culminations, that is

$$\frac{56° \ 47·4′ - 55° \ 11·9′}{2} = 47·8′.$$

This is half an arcminute below the true value for that date in 1984. Because of the pull of the sun and the moon, the earth's axis of rotation slowly precesses like the axis of a spinning top causing the north celestial pole to follow a roughly circular path through the stars in 25 800 years. The pole is now approaching its nearest to Polaris, about 27′ in the year 2095. In Ptolemy's time the distance was 12° while 12 000 years from now, and half way through the next cycle, the distance will be twice 23½°, that is 47°, and the bright star Vega will have taken over as pole star.

A little basic practical work of this kind makes one appreciate the careful results of early observers who generally used hand-graduated circular arcs and open sights in the days before telescopes. In finding the latitude I made use of a calculator which is really unfair, equivalent to using a very accurate table of cosines to find angle A from the lengths of straight sides, and much easier than graduating and centring a circular arc.

Again, the experiment was done in calm conditions on land, very different from the perilous position of a navigator on one of the old sailing ships where an error of a few arcminutes could prove fatal. And yet this particular difficulty was solved by the invention of the sextant, an ingenious instrument developed by John Hadley from an earlier design by Sir Isaac Newton. With it, angles of altitude or of separation could be measured accurately even from the unsteady deck of a ship, using a property of two plane mirrors fixed relative to each other. The sextant brought a new precision to navigation and created a need for good astronomical predictions from the observatories. To conclude this chapter we look at how the instrument works.

In Figure 2.5, P and Q are the mirrors which are attached to a frame not shown. The lower mirror Q is fixed perpendicular to the frame and is half silvered which allows some light from a first object beyond A to pass directly to the observer at B, while the

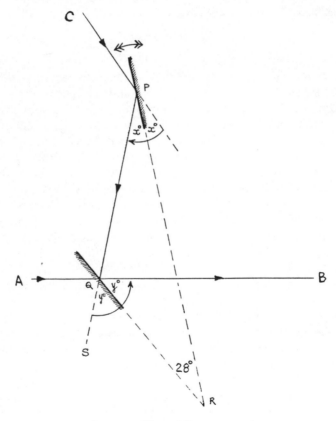

Figure 2.5

Polaris and the finding of latitude

upper mirror P is fully silvered and can be rotated in the manner shown by the double-headed arrows. Light from a second object beyond C is reflected by both mirrors and when P is correctly adjusted this light also emerges towards B. In this way the observer sees both objects brought into coincidence.

In the illustration the angle between the mirrors is set at 28°. Angle SQR being exterior to triangle PQR equals the sum of the interior angles QPR and QRP;

$$y = x + 28 \quad \text{or} \quad y - x = 28.$$

The ray from C is deviated first through $2x°$ one way and then through $2y°$ the other way; the net deviation is

$$2y° - 2x° = 2(y° - x°) = 54°.$$

When the device is tilted x and y will change but their difference $y - x$ will remain equal to 28; and the deviation remains 54° so the images still coincide. Moreover the angle between the two objects being viewed is just twice the angle between the mirrors. A real sextant incorporates a small telescope to help in setting the images on each other, and the movable mirror is linked by an arm to a circular scale.

Newton, towards the end of his own description of his early instrument, describes how the distance of the moon from a star would be found:

By this instrument, the distance of the moon from any fixed star is thus observed: view the star through the eyepiece by the direct light and the moon by the reflexed (or on the contrary) and turn the index till the star touch the limb of the moon, and the index shall show, on the brass limb of the instrument, the distance of the star from the moon's limb; and though the instrument shake by the motion of the ship at sea yet the moon and star will move together as if they really did touch one another in the heavens; so that an observation may be made as exactly at sea as on land.

And by the same instrument may be observed, exactly, the altitudes of the moon and stars, by bringing them to the horizon; and thereby the latitude, and times of observations, may be determined more exactly than by the ways now in use.[1]

For anyone interested in experimenting with a sextant, there are two corrections to the observed altitude of a star using a sea horizon, assuming the instrument itself to be in adjustment.

Refraction. A star at altitude a appears too high by an amount tan $(90° - a)$ arcminutes; approximately true if a exceeds 20°.

Dip of the horizon. From a height h feet above the water, the sea horizon appears depressed below the true horizontal by an amount $\sqrt{h}$ arcminutes. (The use of feet is deliberate – the formula is then so simple.)

For example, Jupiter in the south crosses the observer's meridian at apparent altitude 27° 15' above the sea horizon. The observer is 25 feet above the water and an almanac gives Jupiter's declination as 7° 21' S. The latitude is found as follows.

Correction for refraction = tan 62° 45' arcminutes = 1·94', say 2'.
Correction for dip   = $\sqrt{25}$ arcminutes = 5'.
Corrected altitude   = 27° 15' − 2' − 5' = 27° 08'.
From Figure 2.6 (a view of the celestial sphere from the east)
Latitude = ϕ = 180° − (90° + 7°21' + 27°08')
= 180° − 124°29' = 55° 31' N.

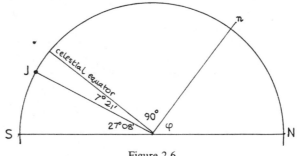

Figure 2.6

Polaris and the finding of latitude

Notes on Chapter 2

Note 2.1. The results of the latitude experiment were:
Upper culmination BC=1104·7 mm, CA=1944·8 mm,
                AB=1920·0 mm;
Lower culmination BC=1148·6 mm, CA=1920·8 mm,
                AB=1920·0 mm.
Angle A was found using the cosine rule for a triangle

$$\cos A = \frac{b^2 + c^2 - a^2}{2bc}$$

in which a=BC, b=CA, c=AB.

A calculator gives 33·209 775° and 34·801 372° for the angles; after rounding they are subtracted from 90° which gives the altitudes.

Note 2.2. If the height above sea level is given in metres, the formula for the angle of dip in arcminutes becomes 1·8 $\sqrt{h}$.

Note 2.3. The extent of atmospheric refraction is quite surprising though it usually goes unnoticed because all parts of the field of view of a telescope are equally affected. For example, Jupiter at an altitude of 30° is seen refracted upwards by tan (90°−30°) arcminutes which is 1·73' or about 100″, nearly three times the apparent diameter of the planet.

3

The earth and the sun: Kepler's first two laws

✳ ✳ ✳

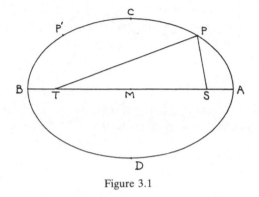

Figure 3.1

Kepler's first law

Johann Kepler (1571–1630) spent many years searching for a simple explanation of the motions of the planets and in 1609 he published the first two of three famous laws that bear his name. The first law has to do with the shape of an orbit and states that each planet revolves round the sun in an ellipse having the sun at one focus.

Figure 3.1 shows an ellipse with its two symmetrically placed foci T and S. The curve can be drawn by making a tight loop of string around pins at T and S and a pencil point at P. The major axis of the ellipse is BA, the minor axis is CD and the centre is M. It is usual to write BM=MA=a so the major axis has length $2a$. The ratio MS/MA is a number e between 0 and 1 and called the eccentricity of the ellipse; in Figure 3.1 $e=\frac{3}{4}$. As the curve becomes more circular the two foci approach each other at M and the eccentricity tends to zero.

Now let the diagram represent the orbit of a planet P around the sun S, the second focus T is just a point in space and is seldom heard of again. When at A, the planet is at *perihelion* nearest the sun, and

$$SA=MA-MS=a-ae=a(1-e).$$

When at B, the planet is at *aphelion* farthest from the sun, and

$$SB=BM+MS=a+ae=a(1+e).$$

The planetary orbits have quite small eccentricities; that of the earth is 0·017 so the ratio of greatest and least distances is $(1+0·017)/(1-0·017)=1·035$, a variation of 3·5 per cent over the year. The apparent diameter of the sun undergoes a similar change, so small that it goes unnoticed.

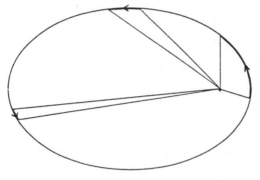

Figure 3.2

Kepler's second law

While Kepler's first law refers to the shape of the orbit, his second law is about the way in which the planet moves. It states that the line joining planet to sun sweeps out equal areas in equal times; in Figure 3.2 three equal areas are shown – the planet would move along the marked arcs in equal times. Obviously it will travel fastest at perihelion and slowest at aphelion. Kepler based this law on careful observations over a long time which can not be emulated here. With a basis of mechanics, as developed soon after Kepler's time, it is easy to prove the second law, and this is done in Note 3.1 at the end of this chapter.

As we are more interested in what can be discovered by experiment, we shall assume the truth of the second law and use it to see the first law in action for the earth, that is, to detect the elliptical nature of its orbit. This may not by itself seem a very important objective, but it so happens that the experimental details are a rehearsal for a later investigation of the outer planets Mars, Jupiter and Saturn which is the subject of Chapter 5.

First, it is shown in Note 3.2 that the second law of equal areas may be written as a formula

$$r^2\omega = k$$

in which

r is the length of the radius from sun to planet,

ω is the angular velocity of this moving radius, measured in units such as radians per second or degrees per day, as convenient, and

k is a constant for the particular orbit.

The helpful feature of this formula is that if r increases by 1 per cent (say from 1 to 1·01), r^2 increases by very nearly 2 per cent (from 1 to 1·0201), so ω must decrease by 2 per cent to maintain the constant right side of the formula. In this way we can deduce small percentage changes in the earth–sun distance from the easier-to-measure changes in angular velocity, and the doubling effect works to our advantage.

But how can we measure the planet earth's angular velocity as seen from the sun? Luckily, the sun appears to us to move round the earth in an orbit exactly the same as the one along which we travel round the sun. When we are at perihelion the sun is at perigee (nearest the earth) and so on. It just depends on the point of view, though this very detail was not understood by those who for many centuries thought that the earth was fixed at the centre of the universe.

The next step is to plan an experiment to follow the angular velocity of the sun at intervals through the year as it progresses through the stars and along the ecliptic. The unit of measurement

The earth and the sun: Kepler's first two laws

is best taken to be degrees or arcminutes per sidereal day, for a reason which will soon be clear. The main requirement is an observing space where static equipment can be left completely undisturbed for several days at a time while the solar system is allowed to spin away between observations. I am fortunate in having a roof space with a skylight opening to the south which has served well both for this experiment and also for investigating the outer planets Mars, Jupiter and Saturn. The general arrangements for timing and so on are much the same for all these experiments and are as follows.

A small hole 2 mm in diameter is firmly fixed near the open skylight and a pinhole image of the sun is received on a flat board about 50 cm square covered with accurately ruled millimetre graph paper. All this is very firmly arranged with the plane of the board perpendicular to the sunbeam and about 3 metres from the hole, at which distance a degree of sky will project on the screen as a length of 5 cm by the one in sixty rule. The image of the sun will be some 2·5 cm across and sharp enough for its position on the graph paper to be read off. Because of the earth's rotation it moves over the screen at an alarming speed – the day is just flying past! But the motion is almost exactly uniform and linear so, even if there are breaks due to cloud, the missing trace can later be completed.

Also present on the screen, though quite invisible in the glare, must be the pinhole images of the stars moving steadily across in an endless panorama. If neither the pinhole nor the screen move, this procession of stars will be repeated in exact detail at intervals of one sidereal day which as we have seen is 86 164·1 seconds of mean solar time. But the sun is the one exception; it travels roughly one degree per day along the ecliptic and so its passage across the graph paper is later each sidereal day by about 4 minutes. By measuring the sun's shift in exactly one – or better in exactly two or up to six sidereal days, one is led to the angular velocity ω. By repeating all this at intervals of a month or so through the year one can find the variations in r from the variations in ω.

Here are some practical details. It must be said emphatically that at no time must one look near the direct sun; to do so is to risk permanent eye damage. All one does is to follow the comparatively faint image of the sun on the screen.

Rather than attempting to estimate the coordinates of the sun's centre on the graph paper, it is easier to note the x-coordinate of the tangent line AB and the y-coordinate of the tangent line

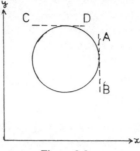

Figure 3.3

CD. Because readings taken a few days apart are to be subtracted, no complication arises from doing this.

For timings in this and other experiments, a small quartz clock which generates an audible tick each second can be strapped to a microphone used with a small portable tape recorder. When starting up, the time shown by the clock is spoken into the machine followed by the observations as they occur, then for safety a final time check. When the tape is played back later, even if it runs at a slightly different speed, there is a background of known ticks which enable the exact time of any observation to be worked out, and of course if there is doubt the tape can be replayed. An effective way of transcribing the tape is to use the lines of $\frac{1}{2}$ cm squared paper at a scale of $\frac{1}{2}$ cm=2 seconds line by line so it can be read like the page of a book.

The original recording is a running commentary on the limb coordinates of the sun; so uniform is the motion that one soon collects enough readings to be reasonably sure of the position at any instant, say to one second of time. To attempt more is rather pointless; there may be slight error or unevenness in the ruled screen and to time to within a second is difficult.

If the sun is clouded over for a while one can just switch off and resume later, not forgetting to put a new time check on the tape. It is well worth having good clusters of readings over a total of 15 minutes; any parts clouded out can easily be interpolated later. All the transcribed information is then plotted on two graphs, x against time and y against time, and all this repeated for the second set of readings a few days later. Of course it is essential not to move the pinhole and screen between the observations.

When the graphs of both observations are finally drawn, two instants are chosen which are an exact number of sidereal days apart, that is, an exact multiple of 86 164·1 seconds, at which times the stellar backgrounds on the screen are identical. Next are found the differences in the sun's x-coordinates and y-coordinates due to its progress against this background. Using Pythagoras' theorem the sun's diagonal movement across the screen is calculated, also the pinhole to screen distances corresponding to the beginning and end of this movement are measured. Finally the angular velocity is found using the cosine formula for a triangle, in the way shown below.

Observations at Edinburgh, 1987 April 20 and April 25

The interval used was five sidereal days from

$$20^d\ 14^h\ 14^m\ 39^s\cdot5 \text{ to } 25^d\ 13^h\ 55^m\ 0^s\cdot0 \text{ universal time (UT)}$$

and chosen because it straddled the reliable parts of the graphs.

$$x \text{ shift} = 122\cdot2\,\text{mm}; \quad y \text{ shift} = 114\cdot0\,\text{mm}$$

$$\text{diagonal shift} = \sqrt{(122\cdot2^2 + 114\cdot0^2)} = 167\cdot1\,\text{mm}$$

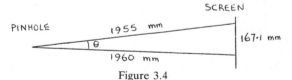

Figure 3.4

pinhole to screen distances 1955 mm and 1960 mm

$$\cos\theta = \frac{1955^2 + 1960^2 - 167\cdot1^2}{2\times1955\times1960}; \quad \theta = 4\cdot8903° \text{ in 5 days.}$$

Do-it-yourself Astronomy

This gives an angular velocity centred on 23 April of 58·68' per sidereal day. The value based on *Astronomical Almanac* information is 58·38' per sidereal day.

Observations of this kind should really be collected for a year and then analysed, but in the time available I obtained just seven sets:

Observation dates		Arcminutes per sidereal day	
		Measured by experiment	Based on almanac
A	1986 October 7, 9	59·4	59·1
B	1986 December 14, 20	60·9	60·9
C	1987 January 2, 5	61·0	61·0
D	1987 January 31, February 6	60·6	60·7
E	1987 March 10, 12	59·5	59·8
F	1987 April 20, 25	58·6	58·4
G	1987 May 29, June 2	57·5	57·3

These are plotted in Figure 3.5 together with the true graph of angular velocity against time of year derived from the *Astro-*

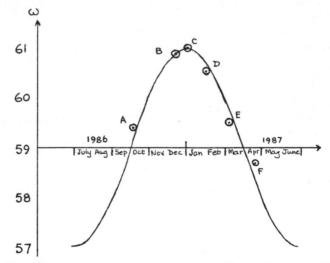

Figure 3.5   The angular velocity of the sun measured in arcminutes per sidereal day: variation during 1986 and 1987

nomical Almanac. The experimental results tend to confirm the annual change in ω of 4 parts in 59, about 6·8 per cent, and earlier we noted that the percentage change in r is just one half of this, namely 3·4 per cent. So the ratio of aphelion and perihelion distances is now $(e+1)/(e-1)=1\cdot034$, from which e, the eccentricity of the earth's orbit, is found to be 0·017.

It so happens that ω is greatest at the season of New Year which is therefore the time of perihelion. By the inverse square law, the earth must receive over 6 per cent more radiation from the sun at perihelion (southern summer) than at aphelion (northern summer), but amongst other things set against this is the high level of reflection from the huge Antarctic ice sheet, so in fact summers in our two hemispheres are not very different.

Notes on Chapter 3

Note 3.1. *Derivation of Kepler's second law from mechanical principles.* In the figure, ABC are in line and AB=BC; also BDEC is a parallelogram. Suppose a body at A is moving with uniform velocity which carries it to B in a first time interval, and which would carry it to C in a second equal time interval, then triangles SAB, SBC have equal bases AB and BC and are equal in area, so the result follows in this case of no central force to S.

If the body when at B receives a sudden impulse towards S and sufficient by itself to carry the body from B to D in one time interval, then by the parallelogram rule for combining vectors the body will in fact move from B to E in the second time interval.

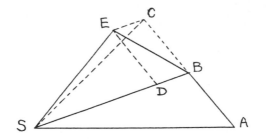

Figure 3.6

But triangles, SBE, SBC are equal in area, having a common base SB and being between parallels EC and SB. Therefore triangles SAB, SBE are equal in area and the result is true for this case of a single impulse and two equal time intervals. By extension it is true for any number of impulses towards S and spaced at equal intervals of time.

A continuous force towards S may be regarded as the limit of a succession of small impulses acting at equally spaced short time intervals, so the result is true in this case also.

(Based on Newton, *Principia* Book 1, Prop. 1, Motte's translation revised by F. Cajori, University of California Press.)

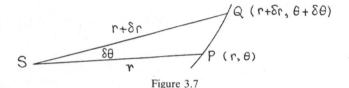

Figure 3.7

Note 3.2. *Derivation of the formula $r^2\omega=k$ from the law of equal areas.* Suppose that in time δt a planet moves from P with polar coordinates (r, θ) to nearby Q with coordinates $(r+\delta r, \theta+\delta\theta)$. The approximately triangular area swept, using the area formula $\frac{1}{2}bc \sin A$ for a triangle, is $\frac{1}{2}r(r+\delta r) . \sin(\delta\theta)$. For a small angle measured in radians, $\sin(\delta\theta)$ can be replaced by $\delta\theta$. So the average rate of area sweeping in time δt is

$$\frac{\text{change of area}}{\text{change of time}} = \frac{1}{\delta t} \times \frac{1}{2} r(r+\delta r) \times \delta\theta.$$

As the small quantities tend to zero the limit of this rate is

$$\frac{1}{2} r^2 \frac{d\theta}{dt} = \frac{1}{2} r^2 \omega$$

where ω is the angular velocity of the line SP in radians per unit time. The second law now assumes the form $\frac{1}{2}r^2\omega=$constant, which may be written $r^2\omega=k$, as in the text of this chapter.

36

4

Venus and Kepler's third law

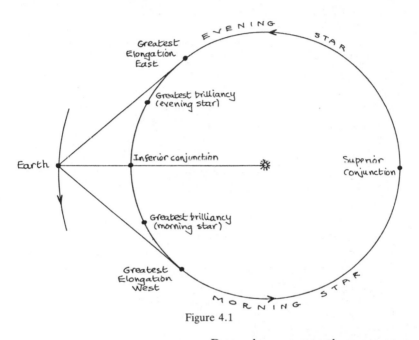

Figure 4.1

Beyond our own earth–moon system, the brilliant planet Venus is the next object to claim attention. Although recent discoveries have shown it to be far from hospitable with its dense, hot and strongly acidic atmosphere, for the naked-eye astronomer it remains a source of unfailing beauty, dominating the scene long after sunset or before sunrise at a favourable elongation. Not only is Venus bright enough to cast distinct shadows when the sky is dark, it can also be followed without optical aid in full daylight for a considerable part of its journey round the sun, and from an extended set of such observations much information can be gained.

37

Do-it-yourself Astronomy

To our modern way of thinking, its progression from evening star to morning star and back again is naturally explained by supposing it to revolve round the sun in an orbit inside that of the earth and roughly coplanar with it, with a periodic time less than our own year. Yet the general acceptance of this idea is less than four hundred years old, being finally settled when Galileo found, with the newly discovered telescope, that the planet could show a phase greater than one-half. This would have been impossible in the old Ptolemaic system of deferents and epicycles which placed Venus always between the earth and the sun. In fairness to the supporters of Ptolemy, they were less concerned with the absolute size or distance of the planets; their main task was to explain and predict the appearances in angular terms and for this the old system worked quite well.

At intervals of 584 days Venus overtakes us in our slower outer orbit; this is the synodic period measured relative to the moving sun–earth line and during which the sequence of events shown in Figure 4.1 takes place.

The further set of diagrams in Figure 4.2 shows an entire synodic period of Venus divided into eight equal parts. A synod is a meeting, and the synodic period is the interval between successive closest approaches or meetings of Venus and the earth. The diagrams also show that in this time Venus makes about 2·6 sidereal revolutions round the sun and relative to the stars while the earth makes about 1·6 sidereal revolutions. Also, the middle diagram (292 days) shows Venus at superior conjunction with the sun, on the far side of the sun as seen from the earth, and of course this happens halfway through the synodic period.

It so happens that five of these synodic periods very nearly equal eight of our years, so the appearances of Venus in our sky recur with little change after an eight-year interval. The table on page 40 gives details of the present cycle and half of the next one, and it will serve as a rough guide for several more cycles if the year numbers are suitably increased.

The altitude depends on the way the ecliptic slants across the sky after sunset or before sunrise and this in turn depends on the time of year. The table shows that a few of the elongations are

38

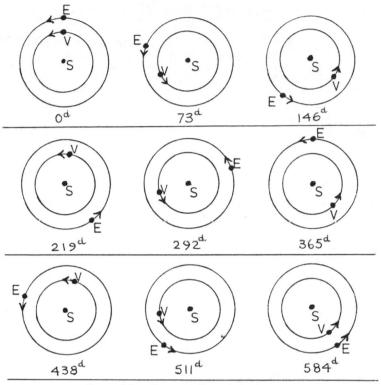

Figure 4.2   One synodic period of Venus

difficult to follow from latitude 56° N. For northern observers conditions are best when Venus is an evening star in spring or a morning star in autumn.

We come now to a practical experiment to follow the movement of the planet through two elongations with the aim of recovering the basic numerical details of the orbit, and also of approaching Kepler's third law which links the motions of two planets. But first, a serious warning that observation of Venus when the sun is above the horizon should be limited to the times around greatest elongation when the planet is 40° or more from the sun. Even so it is dangerous and useless to hunt for Venus in

39

Greatest elongations (about 45° or 46°) of Venus from the sun

Date	East (evening star) or West (morning star)	Altitude at sunrise/sunset at 56° N	at 40° N
1988 April 3	E	39°	45°
1988 August 22	W	32°	40°
1989 November 8	E	6°	19°
1990 March 30	W	9°	20°
1991 June 13	E	22°	34°
1991 November 2	W	34°	42°
1993 January 19	E	28°	39°
1993 June 10	W	12°	25°
1994 August 24	E	6°	18°
1995 January 13	W	16°	29°
1996 April 1	E	40°	45°
1996 August 20	W	31°	39°
1997 November 6	E	6°	19°
1998 March 27	W	10°	21°
1999 June 11	E	23°	35°

a bright hazy sky, the best conditions being after the passage of a cold front when the sky is washed clean and blue. At no time must one stare anywhere near the direct sun; to do so is to risk permanent eye damage. **Safety first** – see Note 4.6.

In Figure 4.3 the base is a square block of wood; the original had sides 20 cm long but it would be better made twice as big. Underneath is some form of universal clamp so the block can be set in any position. DC is a tube or inverted piece of channel set along most of one diagonal of the block; end C is closed except for a small hole to admit sunlight. Just inside end D is a pin which is lined up permanently with another pin A when viewed through the hole C. Once this alignment is made, end D is closed with a piece of tracing paper forming a little screen. The arm pivoted at E carries another pin at its other end B. The pivot is moved from E to the alternative position F when Venus is a morning star.

Venus and Kepler's third law

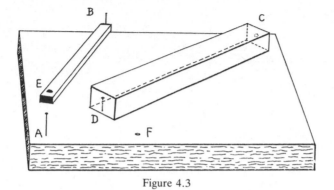

Figure 4.3

The device is turned on its universal mounting into the plane containing Venus, the sun and the observer and is adjusted so the sun shines through the hole forming a small image on the paper screen. This image is easily seen *from the side*; there is no need to look near the sun itself. When the image is bisected by the dark shadow of the pin just inside the tube, ADC will be directed sunwards. Next the hinged arm is moved until the pins A and B are in line with Venus – it is best to stand well back to see this. The planet's elongation from the sun is angle BAC and is found after measuring the lengths of the sides of triangle ABC by using the cosine rule for a triangle with known sides, as was done in the latitude experiment in Chapter 2. If the arm is left undisturbed it helps in finding Venus in daylight on the next occasion. If Venus does not show up within seconds when its direction is known within a degree or so, give up and wait for a better sky.

The experiment began in December 1984 when Venus was a brilliant Christmas evening star rapidly approaching the earth as its elongation increased.

Figure 4.4 summarises the observations during 1985 with a gap around inferior conjunction and other unintended gaps during the western elongation due to an exceptionally cloudy spring and summer in Edinburgh. The greatest eastern elongation came out as 46·8° on 21 January (in fact 47·0° on 22 January), the greatest western elongation as 45·6° on 12 June (in fact 45·8° on 12 June) and inferior conjunction on 5 April (in fact 3 April).

41

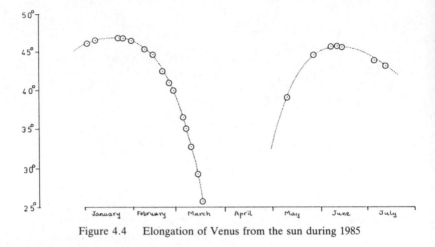

Figure 4.4   Elongation of Venus from the sun during 1985

There is an obvious departure from symmetry in the two branches of the diagram; this must mean that the orbits of Venus or the earth or both are not circular. In fact it is the earth's orbit that is mainly responsible; as we have seen, our distance from the sun is about 3·4 per cent less at New Year than at the beginning of July. The effect on the elongation of Venus is noticeable because its orbit approaches us quite closely, and this particular pair of elongations fell in just the right months to show the difference well. But any adjustment must wait until we have an idea of the size of the orbit of Venus.

The sum of the angles of greatest elongation obtained experimentally was 92·4°, and if once more we assume concentric circular orbits for the two planets, then the angles of quadrilateral EVSW in Figure 4.5 are as shown; in particular angle VSW is 87·6°. As Venus moved from V to W in 142 days this is a fraction 87·6/360 of a complete synodic period; the period itself is then $(360/87·6)\times142$ days which is 584 days and which by luck happens to be correct.

Using Figure 4.6 which is merely half of Figure 4.5 divided on the assumption of symmetry, we can estimate the radius of the orbit of Venus as a fraction of the mean radius of the earth's orbit,

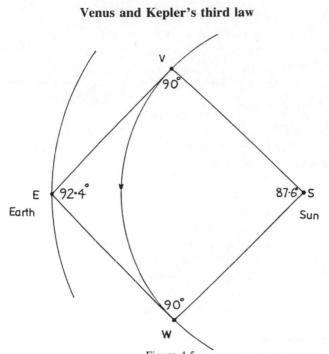

Figure 4.5

an important standard known as the astronomical unit (AU). In Figure 4.6, ES=1 AU and angle EVS=90°, hence

$$VS=1\times\sin 46\cdot2°=0\cdot722 \text{ AU}.$$

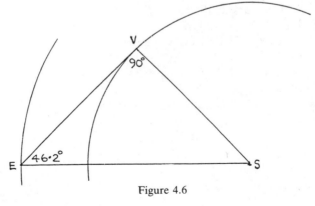

Figure 4.6

Do-it-yourself Astronomy

With this result we can consider the effect of the eccentricity of the earth's orbit on the apparent motion of Venus, and this is done in Note 4.1.

At this stage we have the relative radii of the orbits of Venus and the earth, namely 0·722 AU and 1 AU, also the synodic period of Venus which is 584 days and of course the sidereal period of the earth which is 365·25 days. The next step is to look for a connecting link between these numbers, something that will extend to the other planets of the solar system. As the synodic period is linked to the earth's motion it seems more promising to calculate instead the sidereal period of Venus which, being based on the wider background of the stars, will be more likely to fit into a general scheme of planetary motions.

In the historical development of the subject, the search for a unifying principle continued to occupy Kepler after his success with the first two laws which deal with a single orbit. His work was extensive, covering philosophy and religion as well as science, and was published in 1619 under the title *Harmonices Mundi*. Just one powerful and elegant item from it is remembered nowadays and is known to us as the third law of planetary motion. It states that the squares of the periodic times of the planets round the sun are proportional to the cubes of their mean distances from the sun. The mean or average distance of points on the orbit ellipse from the sun equals the semi-major axis which was denoted by a in Chapter 3 (see Note 4.2).

To find the sidereal period of Venus, note that the synodic period of 584 days implies that in one day the sun–Venus line turns through 1/584 revolution relative to the sun–earth line which in turn rotates through 1/365·25 revolution relative to the stars (time here is in solar days, not sidereal days). As Venus is over-hauling the earth, the sun–Venus line turns in one day through $(1/584)+(1/365\cdot25)$ revolution relative to the stars; this is 0·004 45 or 1/224·7 of a revolution, giving a sidereal period of 224·7 days. (Just why the stars come into it is an interesting point to which we shall return later. Suppose we had lived in a slightly opaque part of the galaxy where the planets were visible but no stars at all; how would we have explained the strange motions in the sky?)

Venus and Kepler's third law

The key figures are now:

Planet	Sidereal period t (days)	Distance from sun a (AU)
Venus	224·7	0·722
Earth	365·25	1

Let us test some likely laws of variation, any of which would make the periodic time increase with increasing distance from the sun.

Try t proportional to a; $\dfrac{224\cdot7}{365\cdot25} = \dfrac{0\cdot722}{1}$; $0\cdot615 = 0\cdot722$

(right side too great)

Try t proportional to a^2; $\dfrac{224\cdot7}{365\cdot25} = \dfrac{0\cdot722^2}{1^2}$; $0\cdot615 = 0\cdot521$

(right side too small)

Try t proportional to $a^{3/2}$, that is, t^2 proportional to a^3

$$\frac{224\cdot7^2}{365\cdot25^2} = \frac{0\cdot722^3}{1^3}; \ 0\cdot378 = 0\cdot376$$

This is so nearly true that we shall regard it as supporting the third law which we have tried to approach through observation, as did Kepler. The result happens to be consistent with the inverse square law of gravitation (Note 6.3), but Kepler's practical rule preceded Newton's exact theory by more than half a century.

Although a precarious leap from our narrow base of observations, we shall assume that the third law applies generally through the solar system and in later chapters see the consequences for the other bright planets, using some more original observations. But before leaving Venus there are a few other details which can reasonably be inferred. As the orbit radius is 0·722 AU, Figure 4.7(a) can be constructed to scale. From the drawing or by trigonometry can be derived any number of distances such as EV_1, EV_2, . . . , EV_5 and from the associated angles EV_1S, EV_2S, . . . , EV_5S can be found the planet's phase by the construction of Figure 4.7(b). Because apparent size is inversely

Do-it-yourself Astronomy

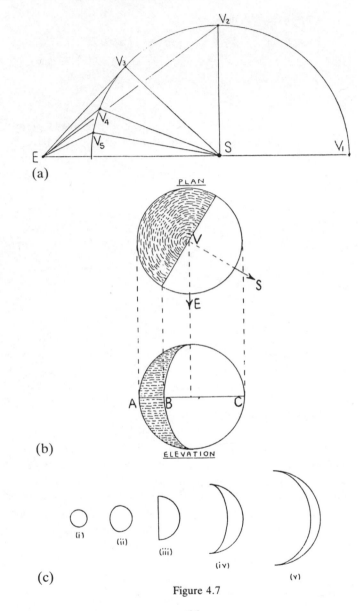

(a)

PLAN

ELEVATION

(b)

(c)

(i)   (ii)   (iii)   (iv)   (v)

Figure 4.7

46

proportional to distance from the earth, the sequence of apparitions in Figure 4.7(c) can now be constructed and verified by telescopic observation. If we suppose that the planet's intense and steady white light comes from perpetual cloud cover over a spherical surface, the brightness of the planet will be nearly proportional to the plane areas of these figures; these can be found because the area unshaded in Figure 4.7(b) is just the fraction BC/AC of the complete circular area. As increasing distance compensates to some extent for change of phase, the variation in the brightness of Venus is not great, just as we observe.

Notes on Chapter 4

Note 4.1. *Effect of the eccentricity of the earth's orbit.* If in Figure 4.5 we use the observed eastern elongation of 46·8° on 21 January, then

$$VS=ES \sin 46·8°; \quad VS=0·729 \ ES.$$

Likewise taking the western elongation to be 45·6° on 12 June,

$$VS=ES \sin 45·6°; \quad VS=0·714 \ ES.$$

But, from Chapter 3,

ES in late January is near $(1-0·016)$ AU$=0·984$ AU
and ES in mid June   is near $(1+0·016)$ AU$=1·016$ AU.

The corresponding values of VS are then 0·717 AU and 0·725 AU; the change being 8 parts in 720 or 1·1 per cent, centred near 0·72 AU. Without knowing the perihelion date of Venus no more can be said, but there seems a chance that the eccentricity of the orbit of Venus is small, maybe smaller than that of the earth's orbit.

In fact the perihelion date was 24 February and the aphelion date was 16 June so the radii VS given above were near the greatest and least possible and the eccentricity is indeed small, well under 0·01.

Note 4.2. *The mean distance of points from the focus is the semi-major axis.* Referring back to Figure 3.1 let P and P' be points symmetrical about the minor axis CD.

Then PS+P'S=PS+PT=AS+AT=AS+BS=2a.

So the mean distance of the pair of points P and P' from S is (PS+P'S)/2 which equals a, and likewise for all such symmetrical pairs of points into which the ellipse can be divided.

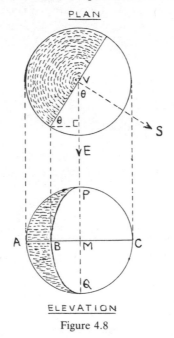

Figure 4.8

Note 4.3. *The phase of a planet.* This is the fraction of the apparent disc that is illuminated. In Figure 4.8. the angle sun–Venus–earth is θ, and if r is the radius of the planet then BM=$r \cos \theta$.

The area of the semi-circle PCQM is $\frac{1}{2}\pi r^2$.

The area of the semi-ellipse PBQM is $\frac{1}{2}\pi . PM.BM = \frac{1}{2}\pi r^2 \cos \theta$.

The phase is therefore $\dfrac{\frac{1}{2}\pi r^2 + \frac{1}{2}\pi r^2 \cos \theta}{\pi r^2} = \dfrac{1 + \cos \theta}{2}$

and varies between 0 and 1.

Venus and Kepler's third law

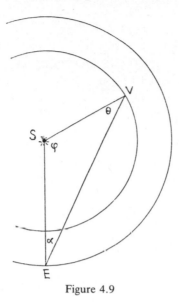

Figure 4.9

Note 4.4. *The elongations of Venus during a synodic period.* In Figure 4.9, SE=1 AU and SV=k AU where k is close to 0·72.

From triangle SVE $\quad\dfrac{\sin \alpha}{k} = \dfrac{\sin \theta}{1};\qquad \sin \alpha = k \sin \theta.$

But $\sin \theta = \sin [180° - (\phi + \alpha)] = \sin (\phi + \alpha)$
$$= \sin \phi \cos \alpha + \cos \phi \sin \alpha,$$
$$\sin \alpha = k \sin \phi \cos \alpha + k \cos \phi \sin \alpha,$$
$$\sin \alpha (1 - k \cos \phi) = k \sin \phi \cos \alpha,$$

$$\tan \alpha = \frac{k \sin \phi}{1 - k \cos \phi}.$$

This gives the elongation α of the planet from the sun in terms of ϕ which for a circular orbit increases uniformly with time.

The following program in BASIC will give the elongation of the planet at daily intervals through a synodic period of 584 days, using $k = 0·72$. The display must be held for inspection by using a suitable key.

```
10 K=0·72
20 FOR T=1 TO 584
30 P=T*2*3·142/584
40 A=ATN(K*SIN(P)/(1−K*COS(P)))
50 PRINT A*180/3·142, T
60 NEXT T
```
(Some computers use ATN instead of ATN.)

Note 4.5. Here is a problem on the subject of Kepler's third law, also with a reference to the factors which affect the brightness of planets, a topic which occurs again in Chapter 7.

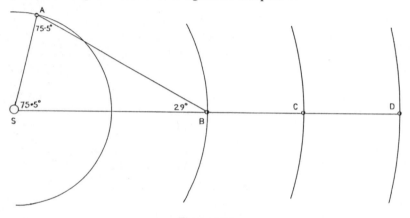

Figure 4.10

Figure 4.10 represents a star with four remarkable planets A, B, C and D which not only have the same size and albedo (reflection factor) but the radii of their circular orbits are in ratio 1:2:3:4. All four planets revolve round the star in the same sense. The only astronomers in the system live on B and they are excited about the alignment of B, C and D as shown in the figure. Show that this event, with C and D in opposition seen from B, last happened 94·36 B-years ago, give or take about half a degree in the sky. Show also that D seen from B appears 9/64 as bright as C seen from B.

The inferior planet A happens to be the same distance from

Venus and Kepler's third law

B as is D; show that its elongation from the star is 29° when seen from B and that it appears 1·4 times as bright as C.

The solution follows – for those who prefer to read it!

Let the unit of time be the B-year, the sidereal period of B round its star relative to the remote background. Then the sidereal periods of C and D are found thus:

$$\frac{t_C^2}{t_B^2} = \frac{3^3}{2^3}; \quad \frac{t_C^2}{1} = \frac{27}{8}; \quad t_C = 1\cdot837\,117\,3;$$

$$\frac{t_D^2}{t_B^2} = \frac{4^3}{2^3}; \quad \frac{t_D^2}{1} = \frac{64}{8}; \quad t_D = 2\cdot828\,427\,1.$$

In 1 year B makes 1 sidereal revolution and C makes 1/1·837.. sidereal revolutions. In 1 year the line SC turns relative to the line SB through $1-(1/1\cdot837..)$ revolutions which is 0·455 668 9 revolution. So in 1/0·455 668 9 years or 2·194 575 9 years the line SC turns once relative to SB; this is the synodic period of C seen from B and is the interval between successive oppositions. Likewise the synodic period of D seen from B is found to be 1/0·646 446 6=1·546 918 2 years. The problem is solved by a time interval which is as nearly as possible an exact multiple of both synodic periods. Form the fraction

$$\frac{2\cdot194\,575\,9}{1\cdot546\,918\,2} = 1\cdot418\,676\,1$$

and with the aid of a personal computer list its multiples; it will be found that 43×1·418 676 1=61 very nearly, from which it follows that 43 synodic periods of C practically equal 61 synodic periods of D, each being a total of 94·36 B-years.

Seen from B, D is twice as far away as C so its disc appears to have one-quarter of the area of the disc of C. Since SC/SD= 3/4, the intensity of illumination at D is only 9/16 of that at C by the inverse square law. Combining the two effects of apparent size and intensity of illumination, D has only (1/4)×(9/16)=9/64 of the brightness of C.

As the sides of triangle ASB are 1, 2 and 2 units it follows from the cosine rule that

$$\cos ABS = \frac{2^2 + 2^2 - 1^2}{2 \times 2 \times 2} = \frac{7}{8}; \quad \text{angle } ABS = 29°.$$

Because AB is twice CB, A too has an apparent diameter one-half that of C. But $SA/SC = 1/3$ so the intensity of illumination at A is 9 times that at C. If A were seen full phase from B it would appear $9 \times \frac{1}{4} = 2 \cdot 25$ times as bright as C. But this has to be reduced to allow for the phase of A. From the isosceles triangle SAB, angle $SAB = 75 \cdot 5°$ and the phase, a little over one-half, is $\frac{1}{2}(1 + \cos 75 \cdot 5°) = 0 \cdot 625$. So A is $0 \cdot 625 \times 2 \cdot 25 = 1 \cdot 4$ times as bright as C if we ignore the slight effect of limb darkening.

A program in BASIC for the above listing is:

```
10  A=2·1945759/1·5469182
20  N=1
30  PRINT N*A, N
40  N=N+1
50  GOTO 30
```

Note 4.6. Measurement of Venus should only be attempted when the planet is well away from the sun near greatest elongation and the sun is near the horizon, so the sky is dark enough for Venus to be seen almost at once. In the experiment described, the image on the tracing paper screen *must* be viewed from the side, so the eye is never directed near the sun.

5

Saturn, Jupiter and Mars
✳ ✳ ✳

We now continue the journey outwards from the sun and attempt by experiments spanning just a few days to find the distances of the easily visible superior or outer planets Mars, Jupiter and Saturn. The main theoretical prop for doing this is Kepler's third law for which we now have at least some evidence from the measures of Venus in relation to the earth.

Quite early in the seventeenth century there began to be observational evidence in plenty. Even before Kepler announced his result in 1619, Galileo directed the newly invented telescope towards Jupiter, revealing the four great satellites like a solar system in miniature. In due course the periodic times and relative distances from the planet were found to fit the third law with perfect accuracy. At the same time a number of scientists suspected that the law of force which accounted for Kepler's results was the inverse square law, and this was eventually proved and set out in careful detail by Sir Isaac Newton in his *Principia*. The easy case of circular orbits is dealt with in Notes 6.1–6.3.

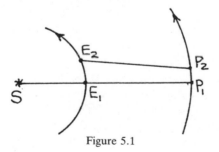

Figure 5.1

Returning to the three superior planets, let Figure 5.1 represent any one of them when nearest the earth and at opposition to the sun. If it were to travel in its orbit at the same speed as the

Do-it-yourself Astronomy

earth, it would at that moment appear stationary against the background of stars, like one aircraft seen from another flying in formation on a parallel course. In fact a superior planet has a lower orbital speed than the earth as we shall soon see, and for a while around opposition it has a reversed or retrograde motion through the stars from east to west as seen from the earth. Eventually it resumes its direct slow progress round the zodiac from west to east until the approach of the next opposition.

In the diagram S is the sun, E_1, P_1 are the earth and planet at opposition and E_2, P_2 are their positions one sidereal day later.

Let the solar distances be $SE_1=1$ AU and $SP_1=a$ AU and the orbit times 1 year and t years. The circumferences are $2\pi\times1$ AU and $2\pi\times a$ AU so the speeds in orbit, using speed=distance/time, are

$$2\pi \text{ AU per year and } \frac{2\pi a}{t} \text{ AU per year.}$$

But by Kepler's law,

$$\frac{t^2}{1^2} = \frac{a^3}{1^3} \quad \text{or} \quad t=a^{3/2}$$

so the speeds may be written 2π AU per year for the earth and $\frac{2\pi a}{a^{3/2}} = \frac{2\pi}{a^{1/2}}$ AU per year for the planet. This shows that the speed of the planet (always supposing circular orbits) varies inversely as the square root of its distance from the sun; at four times the distance the speed is halved and so on. Thus P_1P_2 is less than E_1E_2 and the observed relative motion is indeed retrograde.

We must take this a little further to obtain a useful formula for measuring up the planets: a kind of range-finding device.

Referring again to the figure, the angle in the sky through which P retrogrades in one sidereal day when seen from E is just the angle between the lines E_1P_1 and E_2P_2, which is

$$\frac{(E_1E_2-P_1P_2)}{E_1P_1} \text{ radians.}$$

54

Saturn, Jupiter and Mars

In the short time of one sidereal day or 1/366 year, and using distance=speed×time for each body,

$$E_1E_2=2\pi\times\frac{1}{366} \quad \text{and} \quad P_1P_2=\frac{2\pi}{a^{1/2}}\times\frac{1}{366} \text{ also } E_1P_1=a-1,$$

all measured in AU. Making these three substitutions, the angle in a day is

$$\frac{2\pi}{366}\left(\frac{1-a^{-1/2}}{a-1}\right) \text{ radians, or } \frac{360}{366}\left(\frac{1-a^{-1/2}}{a-1}\right) \text{ degrees.}$$

Now $a-1=(a^{1/2}-1)(a^{1/2}+1)$ and $1-a^{-1/2}=a^{-1/2}(a^{1/2}-1)$ so the angle in a day is

$$\frac{360}{366}\left(\frac{(a^{1/2}-1)\,a^{-1/2}}{(a^{1/2}-1)(a^{1/2}+1)}\right) = \frac{360}{366}\left(\frac{1}{a+a^{1/2}}\right).$$

So finally the retrograde motion at opposition is

$$\frac{360}{366}\left(\frac{1}{a+\sqrt{a}}\right) \text{ degrees per day.}$$

This is close to $1/(a+\sqrt{a})$ degrees per day which is easily remembered. For example if a planet were 9 AU from the sun and therefore 8 AU from the earth at opposition, the daily rate would be about $\dfrac{1}{9+\sqrt{9}} = \dfrac{1}{12}$ degree per day=5 arcminutes per day.

In reverse, if we measure the daily rate, the distance of the planet can be estimated. We just have to hope that the orbits are not so eccentric that large errors are introduced. The rate does not change rapidly at opposition so the measures can be spread over a number of days; thus some interruption by cloud can be survived and indeed the longer time span can help the result by effectively widening the range-finding base.

With the above formula I was ready to observe the three planets, all of which came to opposition in the summer months of 1986. The method has something in common with the measurements of the sun described in Chapter 3; the star background can be relied on to appear in exactly the same way at intervals of one sidereal day, but the planet moves through the stars and crosses a fixed line at intervals which are less than one sidereal day. This time difference can be converted into an angle difference and inserted in the formula to give the planet's distance.

It was convenient to make the observations through the same loft window as before, where everything could be left undisturbed over the necessary interval of some days. Luckily all three planets had southern declinations so their altitudes were not great even when crossing the meridian; this made it easier to set up the basic sighting device. At the south or window end a straight metal edge was fixed exactly perpendicular to the direction of approach of the planet as it travelled nearly horizontally across the sky at about 1 degree in 4 minutes owing to the earth's rotation. Some 3 metres to the north was fixed a plate with a row of holes about 4 mm diameter, through any of which the disappearance of the planet behind the straight edge could be observed and timed. The spacing of the holes is not critical; the idea is to collect a good supply of observations so in the event of patchy cloud at least some will survive to be matched with others made through the same holes on another occasion. Also, plenty of results can be averaged with some chance of increased accuracy. In proceeding to a result two other factors have to be allowed for, first the planet's zodiacal angle of climb mentioned in Chapter 1 and second its (for me) southern declination which slows its apparent speed across the sky.

The best way of explaining the details is simply to work through my results, first for Saturn which was the easiest planet, being far south near position J in Figure 1.4 and therefore with a very small zodiacal descent angle.

Observations from Edinburgh at latitude 56° N.
Altitude of celestial equator in the south $=90°-56°=34°$.
Altitude of Saturn when in the south $=15°$.
Declination of Saturn $=15°-34°=-19°=19°$ S.
Observations on 1986 May 22 and May 28, judged to be near opposition, the planet being in the south at local midnight.
Interval of 6 sidereal days $=6\times86\,164^s\cdot1\ =516\,984^s\cdot6$

Saturn, Jupiter and Mars

Mean of 12 timings of the interval between disappearances of
Saturn behind the eclipsing bar $\quad\quad = 516\,871^{s}{\cdot}7$

Time difference $\quad\quad\quad\quad\quad\quad\quad = \quad 112^{s}{\cdot}9$

This is the change in the planet's right ascension in 6 sidereal days
and is equal to a rate of $18^{s}{\cdot}8$ per day; we now convert to angle.

In 4^{s} a point on the equator moves 1 arcminute.

In 4^{s} a point at declination 19° S moves $(1\times\cos 19°)=0{\cdot}9455$
arcminute.

In $18^{s}{\cdot}8$ such a point moves

$$0{\cdot}9455\times \frac{18{\cdot}8}{4} = 4{\cdot}4438 \text{ arcminute} = 0{\cdot}074\,06°,$$

as seen from the centre of the celestial sphere.

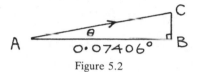

Figure 5.2

This is the shift AB in the diagram. If the zodiacal descent angle
θ corresponding to 19° S were even as great as 5°, which is unlikely,
the cosine of 5° is 0·996 so AC would differ from AB by about
$\frac{1}{2}$ per cent which in this experiment can be neglected. So we take
the daily retrograde motion AC of the planet to equal AB.

Finally, the distance of Saturn from the sun is given by

$$0{\cdot}074\,06 = \frac{360}{366}\left(\frac{1}{a+\sqrt{a}}\right) \quad \text{or} \quad a+\sqrt{a}=13{\cdot}281.$$

This is a quadratic equation in $\sqrt{a}$ and may be solved by formula
or by a few trials with a calculator; the result is $a=10{\cdot}1$.

The conclusion is that at the 1986 opposition Saturn was
10·1 AU from the sun; in fact the distance was nearer 10·0 AU.

Later in the year Jupiter came to opposition and was observed
in the same way, the main difference being that it was much nearer
the celestial equator than Saturn, somewhere between M and A
in Figure 1.4 and with a zodiacal climb/descent angle which I

estimated to be 21°. Here are the details set out in the same way as before.

Observations from Edinburgh at latitude 56° N.
Altitude of celestial equator in the south=90°−56°=34°.
Estimated altitude of Jupiter in the south=27°.
Declination of Jupiter=27°−34°=−7°=7° S.
Observations on 1986 September 9 and September 15, judged to be near opposition, the planet being in the south at local midnight.

Interval of 6 sidereal days=6×86 164^s·1=516 984^s·6

Mean of 12 timings for Jupiter   =516 807^s·0

(standard deviation 2·7 seconds)

Time difference   =   177^s·6

This is the change in the planet's right ascension in 6 sidereal days and is equal to a rate of 29^s·6 per day; we now convert to angle.

In 4^s a point on the equator moves 1 arcminute.

In 4^s a point at declination 7° S moves (1×cos 7°)=0·9925 arcminute.

In 29^s·6 such a point moves $0·9925 \times \dfrac{29·6}{4} = 7·345$ arcminute=0·1224°, as seen from the centre of the celestial sphere.

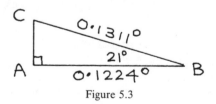

Figure 5.3

This is the shift AB in the diagram. As mentioned above, the zodiacal climb/descent angle was estimated at 21° so the daily displacement of Jupiter was BC=$\dfrac{0·1224}{\cos 21°}$=0·1311°.

Saturn, Jupiter and Mars

Finally, the distance of Jupiter from the sun is given by

$$0\cdot1311 = \frac{360}{366}\left(\frac{1}{a+\sqrt{a}}\right) \quad \text{or} \quad a+\sqrt{a}=7\cdot502.$$

Again a quadratic in $\sqrt{a}$ leads to the solution which is $a=5\cdot22$. The conclusion is that Jupiter was $5\cdot22$ AU from the sun; in fact the distance was $5\cdot0$ AU. Some of the error is due to an underestimate of climb/descent angle which was in fact close to $23°$; the rest is mostly due to the eccentricities of the orbits of Jupiter and the earth.

Finally comes Mars whose complicated motion in the sky occupied Johann Kepler for years; eventually the problem forced him in the direction of his first two laws which solved matters at a stroke. The fact is that the orbit of Mars is both eccentric and comparatively close to that of the earth, so the present method which depends on the displacement difference $E_1E_2-P_1P_2$ and also on the separation E_1P_1 is vulnerable and no great accuracy can be expected.

At this 1986 opposition Mars was at an extreme southerly declination; from Edinburgh it appeared low over hills which often generated cloud. Only one pair of readings emerged but it was reliable and spanned eleven days.

Observations on 1986 July 6 and July 17 near opposition
South altitude of Mars$=7°$;
Declination$=7°-34°=-27°=27°$ S

Interval of 11 sidereal days$=11\times86\,164^s\cdot1 \quad =947\,805^s\cdot1$

Interval between Mars disappearances $\quad =946\,975^s\cdot0$

Time difference $\hspace{6.5cm} = \quad 830^s\cdot1$

Change in right ascension per day$=\dfrac{830^s\cdot1}{11}= \quad 75^s\cdot5$

Angle in sky at $27°$ S

$$=\frac{75\cdot5\times\cos 27°}{4}=16\cdot81 \text{ arcminute}=0\cdot280°.$$

59

Do-it-yourself Astronomy

As with Saturn the zodiacal climb/descent angle is small and no correction is needed. The distance of Mars from the sun is given by

$$0 \cdot 28 = \frac{360}{366} \left(\frac{1}{a + \sqrt{a}} \right) \quad \text{or} \quad a + \sqrt{a} = 3 \cdot 511$$

which leads to $a = 2 \cdot 07$.

The conclusion is that Mars was $2 \cdot 07$ AU from the sun; the correct value was $1 \cdot 41$ AU.

With the exception of Mercury which is close to the sun and difficult to measure without proper instruments, we have now collected results for all the major planets from Venus to Saturn, and with the special formula this has been done over a limited time span as is appropriate for a project of this kind. It is of course a flimsy structure held together by Kepler's third law which was suggested by the motions of Venus and the earth. On the practical side the vital items are the clock backed up by a professional time service, the foot rule or metre scale, a pocket calculator with trigonometric functions and a good supply of graph paper, bits of wood, metal, pinholes and so on.

Here is a summary of results together with correct mean values in brackets. The periods for the outer planets have been obtained from the distance estimates using Kepler's law.

Distances and sidereal periods of the planets

	Distance from the sun (AU)		Sidereal period (days)	
Venus	0·72	(0·72)	225	(225)
Earth	1	(1)	365	(365)
Mars	2·07	(1·52)	1090	(687)
Jupiter	5·22	(5·20)	4360	(4333)
Saturn	10·1	(9·54)	11700	(10759)

These figures give a vivid impression of the scene at the enormous range of Saturn, on what was for centuries thought to be the outer limit of the solar system. The sun at ten times its usual distance would have a barely discernible disc of diameter 3 arc-minutes – it would give practically no heat and already be quite star-like. Jupiter would reach a greatest elongation from the sun of arcsin (5·22/10·1), about 31°, but the earth with its teeming life would pass almost unnoticed being never more than about 6° from the tiny sun, little more than the separation of the Pointer stars in Ursa Major. They at least would look familiar together with the other constellations. How isolated one would feel, looking back at that spark of a sun! How long awaited the return to earth, warmth and light, and no doubt with an increased determination to conserve our vulnerable planet and to live on it more constructively.

The formula $1/(a+\sqrt{a})$ has some other uses. The planet Venus for part of its synodic period has a retrograde motion which is greatest when it passes between the earth and the sun at inferior conjunction – just why this is so may be seen by reversing the roles of E and P in Figure 5.1. Writing $a=0·72$ for Venus, the retrograde rate is $1/(0·72+\sqrt{0·72})=0·638$ degrees per day or 38 arcminutes per day. The planet usually passes a little above or below the sun at inferior conjunction, but very rarely it actually crosses the sun's disc, appearing then as a black dot. As the sun has a direct motion through the stars of about one degree per day, the relative speed of Venus across the background of the sun during such a transit is $38'+60'=98'$ per day, and since the sun's diameter is $32'$ a central transit would last for (32/98) day or $7^h 50^m$.

Only five transits of Venus have so far been seen, those in the years 1639, 1761, 1769, 1874 and 1882, but they have been the centre of much interest. They occur in pairs eight years apart, another effect of that eight-year cycle in the phenomena of Venus; the earlier transit in 1631 was indeed predicted by Kepler but with insufficient accuracy and so was missed. In 1639 the event was correctly predicted by Jeremiah Horrocks (Horrox), curate of Hoole near Preston, England, a young man of genius and great promise who died soon afterwards while still in his early twenties. The day of the transit was a Sunday, and after he had first attended to his church duties he was in time to see the skies clear for half an hour before sunset, revealing the first part of the transit. The only other observer was his friend William Crabtree who at Manchester saw brief glimpses through cloud.[2]

During the seventeenth century it was suggested by the Scottish mathematician James Gregory that transits might be used to investigate further the absolute size of the solar system which is the subject of the next chapter. This idea was taken up by Edmond Halley (1656–1742) whose interest was roused after seeing a transit of Mercury under perfect conditions on the island of St Helena in 1677 (Note 5.3). After his death his ideas and later modifications of them were widely tested at the four Venus transits from 1761 onwards.

The next transit of Venus is on 7 June 2004, though there is a fairly close approach on 10 June 1996 when at 15^h UT Venus will be only 14′ from the sun's limb. Both these events were predicted by Halley to be transits. Because the orbit information available to him was incomplete he was a little off line, but it is remarkable that his timings of both conjunctions, projected from the year 1691 across more than three centuries to dates still in the future, are correct within about 50 minutes.[3] Although not a transit, the 1996 conjunction will be a closer inferior approach than any since the invention of the coronagraph, a means of studying the sun's faint outer envelope without an eclipse. Maybe with such an instrument the black dot of Venus will be detected against the background of the corona.

For a further application of our formula, if we write $a=1$ it gives a daily rate of $1/(1+\sqrt{1})=0.5$ degrees per day. To illustrate this, suppose two spacecraft are in orbit around the sun at a distance from it of just 1 AU, but well clear of the earth. Suppose they are 600 metres apart in line with the sun and proceeding in concentric circular orbits. The outer craft (B) is 'in opposition' as seen from the inner craft (A) and they are far enough apart for their mutual attraction to be negligible. The line AB will rotate $0.5°$ per day in the retrograde sense while the line from the sun to A will rotate $1°$ per day in the direct sense. Relative to the line from the sun to A, the craft B will slip back in a day by $1.5°$ at a range of 600 metres, which is 15 metres by the 1 in 60 rule, and as the days go by the gap will widen as A draws slowly ahead.

Saturn, Jupiter and Mars

Because all this depends ultimately on Kepler's third law, and if we accept that the law holds for a system of satellites circling a planet as well as for a system of planets circling a star, then a numerical change shows us what would happen to two nearby spacecraft in concentric circular orbits around the earth and not far beyond our atmosphere. The orbit time is now about 90 minutes instead of one year, so the line from the inner craft to the earth's centre now turns directly through one degree in a quarter of a minute and the other craft 600 metres farther out will drop back 15 metres as before but in this shorter time of 15 seconds, a rate of just 1 metre per second. When the radial separation is reduced to 6 metres this is still a drift of 1 centimetre per second for concentric circular orbits, enough to make docking quite a tricky operation.

Before leaving the outer planets with which this chapter began, it is worth recalling a remarkable event in 1954 involving Jupiter. Because it was predicted in detail I could not in fairness

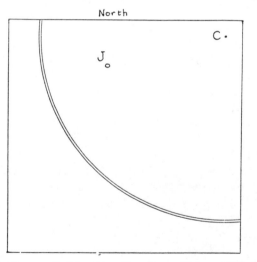

Figure 5.4   Jupiter occulted by the eclipsed sun, 30 June 1954, 12ʰ UT, mid totality near Iceland. C is the centre of sun and of moon; J is Jupiter beyond the sun. Inner arc is the sun, semi-diameter 15′ 45″; Outer arc is the moon, semi-diameter 16′ 05″; diameter of Jupiter 32″

incorporate it in this project, the formula $1/(a+\sqrt{a})$ being a more satisfactory way of tackling the planet's distance and periodic time. But the fact is that on 30 June 1954 there was a total eclipse of the sun which I was fortunate enough to see in Sweden. Jupiter was at superior conjunction on the far side of its orbit and one might have expected to find it near the sun in the dark sky during totality. But there was nothing to be seen for the very good reason that the planet was exactly behind the sun, together with the four major satellites. The earth, moon, sun and Jupiter in that order were then in line, an event unlikely to occur again for something like twenty thousand years (Note 5.2).

Next, moving on just one day less than thirty years to 29 June 1984 we find the earth in practically the same point of its orbit (because of the coincidence of dates) but Jupiter now in opposition, having completed two and a half sidereal revolutions in thirty years which suggests a sidereal period of 12·0 years. In fact the correct period is 11·86 years, the difference being due to the eccentricity of the planet's orbit. In those 30 years Jupiter had been three times round the slow end of the orbit remote from the sun, but only twice round the fast end near perihelion.

That eclipse in 1954 was the most recent one visible from any part of the British Isles. A colleague of mine Michael Lester-Cribb, for many years Director of Music at Fettes College, joined a party travelling by boat to Unst in the Shetland Isles which was to be in the belt of totality. After landing at Baltasound, Michael ran north as fast as he could and managed to get to Haroldswick; although the event was on an astronomical scale every mile northwards meant extra seconds of totality. The weather was generally cloudy with squalls but at the right moment a break in the cloud gave him a perfect view of the whole of the total phase, the sun being at altitude 52°. Three observers in the north-east of the island, Dr Lyne-Pirkis, Mr K. Brierley and Mr G. C. Marlowe saw about three-quarters of totality while a party of twenty on a nearby hill saw nothing. Michael, with the possible exception of someone in an isolated croft, may have been the only person based on the British Isles to see the whole total phase. But he did not see Jupiter!

Saturn, Jupiter and Mars

Notes on Chapter 5

Note 5.1. The question of Jupiter's position at the 1954 eclipse was raised by Mr T. J. E. Palmer at a meeting of the British Astronomical Association in April 1954 and the details confirmed by Dr W. H. Steavenson, the well-known observer of those days (*Journal of the British Astronomical Association*, vol. 64, page 277).

The report of Dr Lyne-Pirkis is from a summary of observations given in the same journal (vol. 65, page 22).

The next total solar eclipse visible from any part of the British Isles will be on 11 August 1999, the belt of totality crossing Cornwall and Devon from west to east. Preliminary details including a map have been given by I. Ridpath in the *Journal of the British Astronomical Association*, vol. 97, page 96 (1987).

Note 5.2. *To estimate the chance of the occultation of Jupiter behind the eclipsed sun.*

1. There is roughly one accessible total eclipse every two years.

2. At the moments of all total eclipses, the positions of Jupiter in relation to the earth–moon–sun line will, over a sufficiently long time, be randomly distributed around the ecliptic.

3. The positions of Jupiter and the sun along the ecliptic (that is, in ecliptic longitude) must agree within about a quarter of a degree, the chance of this being one in 4×360 or 1 in 1440.

4. The orbit of Jupiter is inclined to the plane of the ecliptic at $1 \cdot 3°$; seen from the earth at superior conjunction (beyond the sun) it can appear up to $68'$ from the ecliptic (its maximum ecliptic latitude), this latitude varying sinusoidally round the ecliptic. Because the sun's radius is $16'$, the sun must be fairly near a node of the planet's orbit (where the orbit crosses the ecliptic); if the greatest allowed angular distance is θ, we must have $68 \sin \theta = 16$; $\theta = 13°$. So only 13/90 of superior conjunctions at moments of total eclipse are times when the planet is actually hidden by the sun.

Since $\dfrac{1}{2} \times \dfrac{1}{1440} \times \dfrac{13}{90} = \dfrac{1}{20\,000}$ nearly, the event may be expected only once in about 20 000 years.

Note 5.3. As a young man Halley visited the island of St Helena (15°55' S, 5°44' W) principally to make a catalogue of southern stars, but he also observed the transit of Mercury on 7 November 1677 (New Style Calendar) in truly memorable circumstances from his temporary observatory high above the Atlantic. Not only was the weather perfect throughout the event which lasted about 5^h 12^m, but at the middle of the transit the sun crossed the local meridian only 40 arcminutes from the zenith, with Mercury then a mere 4 arcminutes from the sun's centre. At both start and finish the sun's altitude was about 50°. In 1691 he referred to '. . . that remarkable transit of Mercury which I fully observed at St Helena . . . and whose middle I determined there at 4^m p.m. but at London 28^m p.m. . . .' Again in 1716 he wrote:

While I was making my observations in the island of St Helena, about 40 years since, on the stars round the south pole, I happened to observe, with the utmost care, Mercury passing over the sun's disc; and contrary to expectation, I very accurately obtained, with a good 24-foot telescope, the very moment in which Mercury, entering the sun's limb, seemed to touch it internally, as also that of his going off; forming an angle of internal contact. Hence I discovered the precise quantity of time the whole body of Mercury had then appeared within the sun's disc, and that without an error of one single second of time; for, the thread of solar light intercepted between the obscure limb of the planet, and the bright limb of the sun, though exceedingly slender, affected my sight, and in the twinkling of an eye, both the indenture made on the sun's limb by Mercury entering into it, vanished, and that made by his going off, appeared. On observing this I immediately concluded that the sun's parallax might be duly determined by such observations.[4]

As Mercury has in fact no atmosphere, and in those exceptional conditions, he may have been able to time the instants of ingress and egress 'without an error of one single second'. After the transit the thought came to him that here was a kind of celestial micrometer waiting to be used to measure the size of the solar system with exceptional accuracy. As our formula has shown, a

Venus transit proceeds at a rate of 32′ in 470^m which is about 1″ in 15^s or only one-fifteenth of an arcsecond in that 'single second of time'. On this theme he wrote 'The principal use of these conjunctions is accurately to determine the distance of the sun from the earth, or his parallax. . . . In observing the ingress of Venus into and egress from the sun, the space of time between the moments of the internal contacts may be obtained to a second of time, that is to 1/15 of a second [of arc], by means of an ordinary telescope and a clock that goes accurately for six or eight hours.'

The idea was ingenious but its execution in ordinary climates was full of difficulties; the extensive atmosphere of Venus and the normal state of turbulence in the earth's atmosphere make the timings uncertain by ten seconds or more. But these would be problems for Halley's successors – in 1677 at St Helena the next transit of Venus was still 84 years ahead.

6

The astronomical unit I

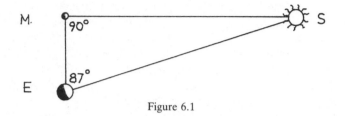

Figure 6.1

It can hardly be disguised that we have so far failed to measure absolutely in miles or kilometres the distance of even one planet; all we have done is to describe a scale model of the solar system based on the mean distance of the earth from the sun, that is, the astronomical unit (AU). Certainly this unit must be many times the distance from the earth to the moon, because when the moon appears half illuminated the angle moon–earth–sun is invariably close to 90°, so moon and earth are receiving nearly parallel rays from the very distant sun.

Aristarchus of Samos (310 B.C.–250 B.C.) is famous as almost the only person in those early days to assert that the earth moved round the sun. Not only this, he also attempted to find the length of the astronomical unit by noting the position of the sun when the moon, seen from earth, appeared half illuminated. He found angle MES in Figure 6.1 to equal 87°. We would proceed to say ME/ES=cos 87°; ES=19·1×EM. But Aristarchus, unable to reach for a pocket calculator as we do, produced an elegant geometrical proof that ES was between 18 and 20 times EM. Although the angle MES is in fact much closer to 90° the method is sound and was important in giving at least an indication of the huge size and distance of the sun.

With technical equipment such as radar or the spectroscope and a knowledge of the velocity of light, it is not difficult to find the astronomical unit. But from our point of view, which is also that of the scientists who followed Kepler, its sheer size makes any form of earth-based triangulation difficult. There is also an interesting theoretical difficulty. Beyond our local earth–moon system *the motions of celestial bodies are such as to conceal their true distances from us.* For all we know from observations such as those of Chapter 3, the sun might be 150 million kilometres away with a diameter of 1·4 million kilometres, or it might be twice as far away and twice as big.

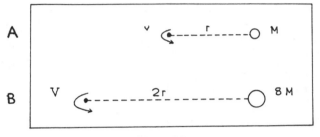

Figure 6.2

To see how this 'concealment principle' comes about, at least for circular orbits, consider the two systems A and B in Figure 6.2. Each consists of a central star with a single small planet in circular orbit round it. The linear dimensions of B are just twice those of A, and the star densities are the same. Then in system A there is a star of mass M with a planet at distance r moving with speed v while in system B there is a star of mass $8M$ (because the volume formula for a sphere involves the cube of the radius) with a planet at distance $2r$ moving with a speed V. It is not difficult to find the link between V and v.

In A, the acceleration of the planet towards the centre of the circular orbit is v^2/r (Note 6.1) and this equals the attractive force exerted by the star on a mass of 1 unit at the planet's distance which is proportional to M/r^2 by the inverse square law.

Thus $\dfrac{v^2}{r} = k \times \dfrac{M}{r^2}$ where k is constant.

69

The corresponding equation for system B is

$$\frac{V^2}{2r} = k \times \frac{8M}{(2r)^2} \text{ in which } k \text{ is as before.}$$

Therefore $V=2v$; the angular velocities $V/2r$ and v/r are equal and so are the periodic times. If B were twice as far away from us as A, not only would the two systems look the same but they would move in the same way: they would be indistinguishable without special equipment. This is why absolute distances in the solar system resist determination by ordinary timing and position measuring.

As all this came to be understood, the problem of the astronomical unit was tackled by what then seemed to be the only way open: long-range surveying using widely spaced stations on earth as a base line, or else a single station moved around by the earth's rotation. Efforts were naturally concentrated on the planets Mars and Venus which approach closest to the earth. In 1672 a favourable opposition of Mars was observed simultaneously by J. Richer in Cayenne, South America, and by G. D. Cassini in Paris. The track of the planet through the stars was followed at each place and later the two records were compared.

Another idea was to observe transits of Venus across the sun, proposed first by Gregory in 1663 and developed in detail by Halley after 1677. It was explained in Note 5.3 that, after his experience at St Helena, Halley hoped that the progress of such a transit could be measured to a small fraction of an arcsecond. He expected the 1761 transit to pass 4′ 15″ south of the sun's centre, making it nearly diametral, and had in mind a calculation based on the shortening of the transit for an observer carried on the sunward side of the spinning earth, and the lengthening for another observer carried the other way through the short northern night, where the transit would begin before sunset and end after the following sunrise. In the event, the track of Venus was farther from the sun's centre as in Figure 6.3, but this enabled another effect to be used: observers in widely differing *latitudes* saw transits of different duration associated with the parallel tracks now cutting the sun's limb at an angle, as shown in the same figure. Although these effects

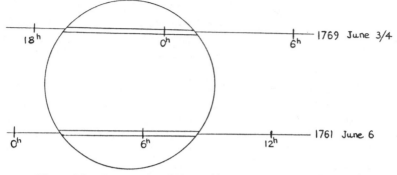

Figure 6.3   Two transits of Venus (times UT, New Style Calendar)

showed up, the difficulty in deciding the moments of contact set a limit on accuracy; it fell far short of what Halley would have wished from his celestial micrometer.

As an example of the moving single station method, Sir David Gill in 1877 observed another favourable opposition of Mars from Ascension Island. Being near the equator he was moved back and forth across a diameter of the earth between successive evening and morning observations of the planet. Again the displacement of Mars against the star background was measured and a good result was eventually obtained.[5]

Inevitably one asks if there is any chance of measuring the astronomical unit by these methods within the terms of this project, without a telescope. The answer seems to be that by Gill's moving station method an effect might just be observed, but it would be so uncertain as to be useless. Moreover, it depends on an almanac prediction which disqualifies it anyway, and it involves Mars whose distance we have measured with the least confidence. It does happen that Mars is in opposition in January 1993 and close to the bright stars Castor and Pollux. By timing disappearances of the trio behind an exactly straight bar set parallel to the earth's axis for a series of evenings (risings in the east) and mornings (settings in the west), a slight residual displacement of Mars in right ascension might be detected. It might amount to one second of time in these latitudes, to be disentangled from other uncertainties of about the same magnitude.

Do-it-yourself Astronomy

So we are faced with a difficult situation, being unable to triangulate effectively and with the barrier of the concealment principle. Fortunately some progress is possible by an indirect approach using the brightness of the planets and this is discussed in the next chapter.

Meanwhile there are some more positive aspects of the concealment principle worth mentioning. In recent years we have become used to seeing artificial satellites moving across the twilight sky like bright points, in near circular orbits just above the atmosphere. It is shown in notes 6.4 and 6.5 that the lowest of such skimming satellites travels at 8 kilometres per second round the earth whose radius we shall find is nearly 6400 kilometres, and the time taken is about 85 minutes.

If an explorer were to land on a spherical non-rotating asteroid with the same density as the earth and a radius of 6·4 km, just one-thousandth that of the earth, then the concealment principle tells us that a good running take-off at 8 metres per second, which is one-thousandth of 8 kilometres per second, would be sufficient to put him into a skimming orbit with the same sidereal period of 85 minutes. Even if his run-off was a shade fast, provided it was at a tangent to the surface he would return to ground level near his starting point where a rope thrown by a colleague, or perhaps a large net, would come in useful.

He would find that acceleration due to surface gravity would be reduced by the same factor of a thousand; on the earth a stone falls 5 metres in one second from rest but our explorer would find that a pebble fell only 5 mm in a second.

On a slightly more serious note and at the other extreme of size, the principle can be used to find the density of the sun compared with that of the earth. Suppose that the sun is shrunk or expanded without change of mass until it has the same density as the earth; this will not affect the motions of the planets moving round it. Suppose also that round this modified sun there moves in a skimming orbit a strange new planet, Vulcan. Having regularised the density, the condition for the concealment principle applies and the sidereal period of Vulcan round the sun will be 85 minutes, or more accurately 5077 seconds.

The astronomical unit I

If Kepler's third law $t^2 = k \cdot r^3$, with which is involved the inverse square law of gravitation (Note 6.3) were to remain accurately true for a skimming satellite or planet, then the radius of Vulcan's orbit which is also the radius of the modified sun, could soon be found. Amazingly, the law does remain accurately true. Provided the central attracting body has spherical symmetry, it attracts external objects just as if it were a particle equal in mass to itself and placed at its own centre (Appendix 1). With this result Kepler's law can safely be applied to Vulcan and the earth in relation to the sun.

For Vulcan, $t=5077$ seconds and r AU is its orbit radius; for the earth, $t=365\times86\,400$ seconds and 1 AU is its orbit radius.

$$\frac{k \times r^3}{k \times 1^3} = \frac{5077^2}{(365\times86\,400)^2}; \quad r = \left(\frac{5077}{365\times86\,400}\right)^{\frac{2}{3}}; \quad r = 2 \cdot 96 \times 10^{-3}.$$

Next, the radius of the real sun seen from our distance of 1 AU is 16 arcminutes $= \dfrac{16}{60} \times \dfrac{\pi}{180}$ radians $= 4 \cdot 654 \times 10^{-3}$ radians subtended at 1 AU, so the real radius is $4 \cdot 654 \times 10^{-3}$ AU.

So, to bring its density up to that of the earth, the radius of the real sun would have to be reduced to a fraction $2 \cdot 96/4 \cdot 654$ of its present value, and its volume to a fraction $(2 \cdot 96/4 \cdot 654)^3 = 0 \cdot 26$ of its present value.

Therefore the density of the sun is $0 \cdot 26$ times the density of the earth.

As a last application of the principle, it happens by chance that the sun and moon appear the same size in the sky, and they are both at distances from us that are many times greater than the diameter of the earth. Each body raises tides in the oceans owing to the change in attractive force across the space of the earth's diameter. If the densities of the sun and moon were equal then the concealment principle would apply and the solar and lunar tides would be of equal strength (Note 6.8). But the solar tide is weaker than the lunar tide, so the sun's density is less than that of the moon which in turn is rather less than that of the earth, though we cannot prove it here.

73

Notes on Chapter 6

Here is a summary of certain connected topics in dynamics, restricted to uniform motion in circular orbits.

Note 6.1. *The acceleration formula* v^2/r. This gives the acceleration of a particle moving with constant speed v in a circle radius r, the angular velocity ω is then v/r radians per unit time. The acceleration is directed to the centre of the circle. Even though the speed is constant the velocity vector is continually changing its direction so there is an acceleration given by $\dfrac{\text{change of velocity vector}}{\text{change of time}}$. Here is a sketch of the proof with only a touch of calculus notation.

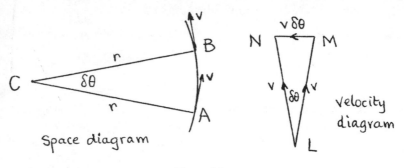

Figure 6.4

$\delta\theta$ is in radians. In time δt the particle moves from A to B and the velocity vector swings from LM to LN, the vector change being MN. The mean acceleration over this short time has magnitude $v \, \delta\theta/\delta t$. In the limit this is

$$v\omega = v \frac{v}{r} = \frac{v^2}{r} = r\omega^2,$$

the direction being that of MN, or from the short arc AB to the centre C.

The astronomical unit I

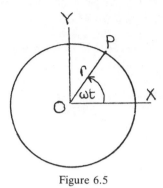

Figure 6.5

Note 6.2. *An alternative proof.* In Figure 6.5 the position vector of the moving point P relative to the centre O is $r=r \begin{pmatrix} \cos \omega t \\ \sin \omega t \end{pmatrix}$. The velocity and acceleration vectors are obtained by differentiating with respect to time: $v=r\omega \begin{pmatrix} -\sin \omega t \\ \cos \omega t \end{pmatrix}$; $a=-r\omega^2 \begin{pmatrix} \cos \omega t \\ \sin \omega t \end{pmatrix}$.

As the vectors within brackets are unit vectors, the magnitudes give $v=r\omega$ along the tangent to the circle and $a=r\omega^2=v^2/r$ towards the centre.

Note 6.3. *Kepler's third law suggests the inverse square law.* If m is the mass of a planet and t is its periodic time in a circular orbit of radius r, the speed being v, then the force of attraction to the centre is given by force=mass×acceleration: $F=m\dfrac{v^2}{r}$.

Also $v=\dfrac{2\pi r}{t}$; $v^2=\dfrac{4\pi^2 r^2}{t^2}$; therefore $F=\dfrac{m4\pi^2 r}{t^2}$.

But by Kepler's law, $t^2=kr^3$ and so $F=\dfrac{m4\pi^2 r}{kr^3}$; $F=m.\dfrac{A}{r^2}$.

According to W. W. Rouse Ball,[6] all this was understood by scientists such as Hooke, Huygens, Halley and Wren by 1684 or earlier. The more complex details for elliptical orbits were filled in by Newton in his *Principia*, 1686.

75

Do-it-yourself Astronomy

Note 6.4. *The orbit speed of a skimming satellite of the earth*. The only force on the satellite is its weight mg towards the earth's centre. If the orbit is circular the motion equation is

$$\text{force}=\text{mass}\times\text{acceleration}; \quad mg=m\frac{v^2}{r}; \quad v^2=rg.$$

Writing $r=6\cdot4\times10^6$ metres and $g=9\cdot8$ metres per second per second,

$v = 7920$ metres per second, about 8 kilometres per second.

Note 6.5. *The periodic time of a skimming satellite*.

$$\text{Time}= \frac{2\pi r}{v} = \frac{2\times\pi\times6\cdot4\times10^6}{7920} = 5077 \text{ seconds}=84\cdot6 \text{ minutes}$$

This is the time relative to the stars. Seen from the earth which is spinning below the orbit, the time will be different.

Note 6.6. *Motion near a large sphere*. An important theorem proved by Newton in 1685 states that the attraction of a large sphere at an external point equals the attraction of an equally massive particle replacing the sphere and located at the sphere's centre. This is discussed separately in Appendix 1. It enables Kepler's law to be extended to objects such as artificial satellites close to the earth; in particular it becomes possible to link the moon's motion with that of a skimming satellite.

Note 6.7. *Transits of Venus across the sun*. Figure 6.3 is based on the computer program given in *Astronomy with your Personal Computer* by P. Duffett-Smith (Cambridge University Press).

Example 1. The same source yielded the following coordinates for Venus and the sun at the next transit on 7–8 June 2004

76

The astronomical unit I

Time (UT)	Venus			Sun's centre			
$8^d \quad 0^h$	5^h	8^m	22^s	5^h	5^m	50^s	RA
	$22°$	$49'$	$12''$	$22°$	$53'$	$11''$	Dec. (N)
$8^d \quad 6^h$	5^h	7^m	42^s	5^h	6^m	52^s	RA
	$22°$	$44'$	$49''$	$22°$	$54'$	$30''$	Dec. (N)
$8^d \quad 12^h$	5^h	7^m	3^s	5^h	7^m	53^s	RA
	$22°$	$40'$	$26''$	$22°$	$55'$	$48''$	Dec. (N)

Sun's semi-diameter 15·8′.

List the coordinates of Venus relative to the sun's centre at the three times and plot the track of the centre of Venus across the sun. Remember to convert the RA change to arcminutes as in Examples 1 and 2 at the end of Chapter 1. Also RA increases to the left of the diagram. (Coordinates in arcminutes: $(35·1, −4.0)$; $(11·5, −9·6)$; $(−11·5, −15·4)$.)

Example 2. Likewise draw the track of the second of this pair of transits, due on 5–6 June 2012.

Time (UT)	Venus			Sun's centre			
$5^d \quad 18^h$	4^h	58^m	50^s	4^h	56^m	51^s	RA
	$22°$	$55'$	$12''$	$22°$	$40'$	$26''$	Dec. (N)
$6^d \quad 0^h$	4^h	58^m	11^s	4^h	57^m	53^s	RA
	$22°$	$50'$	$40''$	$22°$	$41'$	$58''$	Dec. (N)
$6^d \quad 6^h$	4^h	57^m	32^s	4^h	58^m	55^s	RA
	$22°$	$46'$	$8''$	$22°$	$43'$	$29''$	Dec. (N)

Sun's semi-diameter 15·8′
(Coordinates in arcminutes: $(27·4, 14·8)$; $(4·1, 8·7)$; $(−19·1, 2·7)$.)

Note 6.8. *Lunar and solar tides.* Although the pull of the sun is much greater than the pull of the moon, the rates of change with distance are not so different and would be equal if the densities were equal. Using Figure 6.2 with its systems A and B as an example, if the small planet there mentioned is the earth with diameter x assumed to be small compared with r, then first for system A the difference in the attractions on unit masses at the ends of this small diameter is

$$\frac{kM}{r^2} - \frac{kM}{(r+x)^2} = \frac{kM[(r+x)^2-r^2]}{r^2(r+x)^2} = \frac{kM(2rx+x^2)}{r^2(r+x)^2}$$

But x^2 is small alongside $2rx$ and x is small alongside r; the difference is close to

$$\frac{kM.2rx}{r^2.r^2} = \frac{2kMx}{r^3}.$$

When the little planet is transported without change of size to system B, we write $8M$ for M, $2r$ for r, but x remains the same. The difference is now

$$\frac{2k.8M.x}{(2r)^3} = \frac{2kMx}{r^3} \text{ as before.}$$

By a similar argument, if the sun and moon were not only of equal apparent size in the sky but also of equal density, then they would raise equal tides.

7

The astronomical unit II

* * *

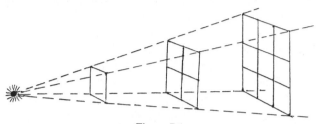

Figure 7.1

We now see that to find the astronomical unit is a difficult problem which really demands accurate measuring equipment. This chapter sketches an attempt to set rough limits between which the value may lie. It is not therefore to be taken too seriously but the factors which affect the brightness of planets may be of interest.

In Figure 7.1 a point source sends light towards parallel windows spaced at equal distances and having all their panes of equal size. The same total light energy which passes through the first window is later spread over 4, 9, ... panes; the intensity of illumination at distances 1, 2, 3, ... from the source is proportional to $1, \frac{1}{4}, \frac{1}{9}, \ldots$ and varies inversely as the square of the distance from the source.

Likewise if a small source of light is viewed from various distances then its apparent brightness varies inversely as the square of the observer's distance. For in the same diagram let a single pane at any distance represent the eye of the observer, and the conclusion follows.

If a number of white circular discs with radii 1, 2, 3, ... units are placed side by side, equally illuminated by a remote source and viewed from a great distance, their brightnesses will be proportional to their areas and therefore varying directly as the squares of their radii.

If a planet is seen from far away to be partly illuminated, showing a phase which is a fraction of the complete disc, then the brightness varies directly as the phase. This is only approximate, thus the moon at first quarter is less than half as bright as a full moon at the same distance. But the statement is nearer the truth for a sphere with a matt white surface.

The brightness of a planet also varies according to its albedo or reflecting power, though in a complicated way depending on the nature of the surface and on the angle of illumination when the planet shows a phase. When the phase is full the planet may brighten a little owing to a hint of specular or mirror-like reflection; likewise, when a planet such as Venus shows just a thin crescent, the light from it has been reflected after almost a grazing incidence which again causes a specular effect. In this way the mean brightness per square arcsecond of a planetary disc can vary with phase. But a cloudy surface will have a much greater albedo than a barren rocky one, and this distinction is really all we need.

Putting this together, the brightness of a planet varies directly as the albedo (a), directly as the phase (p), directly as the square of the radius (r), inversely as the square of the planet's distance from the sun (s) and inversely as the square of its distance from the earth (e).

This is summed up in the formula for brightness (b):

$$b = k \cdot \frac{apr^2}{s^2e^2} \text{ in which } k \text{ is constant.}$$

Here is an illustration, using true values, of the effect of some of these factors. The distance between the earth and Venus when they are closest happens to be about 100 times the distance from the earth to the moon. From the dark side of Venus the earth would dominate the night sky, appearing full phase and at opposition, and with the full moon alongside. The moon's brightness at that distance would be reduced to a fraction $1/100^2$ or $1/10\,000$ of what we see. Because five magnitudes on the stellar scale correspond to a brightness ratio of $100:1$ the reduction would be ten magnitudes from the usual $-12 \cdot 7$ to $-2 \cdot 7$, making it about as bright as Jupiter. Nearby would be the full earth having four times the moon's radius and sixteen times the disc area. Moreover the air and cloud cover give it at least three times the moon's albedo, so both factors would make it some fifty times, or $4 \cdot 3$ magnitudes, brighter than the moon alongside, and approaching magnitude -7 which is far brighter than Venus ever appears to us on earth. Depending on the time of month the separation of the moon and the earth might be anything up to arctan $(1/100)$ which is $0 \cdot 57°$, roughly the moon's diameter as we see it.

What a splendid sight our binary system would be at that range, with or without a telescope. At first, how little would a stranger approaching our two worlds suspect the almost total difference between them: the smaller darker moon, unable through its weak gravity to retain an atmosphere; barren, stony and unchanged through most of its long existence, and the larger brilliant earth, highly evolved, teeming with life and in this most recent flash of its history being changed for ever by its dominant species.

But we must return to our formula and the search for the astronomical unit. Let suffixes 1 and 2 refer to two planets so that

$$b_1 = k \cdot \frac{a_1 p_1 r_1^2}{s_1^2 e_1^2} \text{ and } b_2 = k \cdot \frac{a_2 p_2 r_2^2}{s_2^2 e_2^2}$$

On eliminating k,
$$\frac{r_1^2}{r_2^2} = \frac{a_2}{a_1} \times \frac{p_2}{p_1} \times \frac{s_1^2}{s_2^2} \times \frac{e_1^2}{e_2^2} \times \frac{b_1}{b_2}.$$

This new result enables the radii of two planets to be compared. For a first application let us examine Venus and Jupiter which were three degrees apart in the dawn sky on 30 January 1984. With the aid of a filter made of a pile of tinted perspex plates previously calibrated, I made Venus to be five times brighter than Jupiter, by reducing the stronger light until it matched the weaker. Next, the continuous brilliant light of both planets suggests that they are cloud covered and with roughly equal albedos, so $a_1/a_2 = 1$. Finally, the phases and distances in AU are found by drawing based on the results of previous chapters.

Do-it-yourself Astronomy

Here is a summary of the data, with rounded true values given in brackets.

Planet	Albedo a	Phase p	Solar distance s (AU)	Earth distance e (AU)	Brightness b
Jupiter	—	1·0	5·22	5·96	1
	(0·52)	(1·0)	(5·28)	(6·03)	(1)
Venus	—	0·81	0·72	1·28	5·2
	(0·49)	(0·82)	(0·72)	(1·28)	(6·9)

(The albedo of Venus takes account of phase effects.)

The formula gives $\dfrac{r_1^2}{r_2^2} = \dfrac{0\cdot81}{1} \times \dfrac{5\cdot22^2}{0\cdot72^2} \times \dfrac{5\cdot96^2}{1\cdot28^2} \times \dfrac{1}{5\cdot2}; \quad \dfrac{r_1}{r_2} = 13\cdot3.$

So Jupiter is enormous compared with Venus, having according to our figures 13 or 14 times the radius. (In fact the ratio of radii is 11·8:1. The figures in brackets will give 11·5:1, the difference is due to many rounded decimals being combined.) As Jupiter at its nearest (5·22−1=4·22 AU) is 15 times as far away as Venus at its nearest (1−0·72=0·28 AU), the two will appear at such times about equal in size. This is consistent with other indications: Venus as a thin crescent is so bright that it can hardly be less than one arcminute in diameter, while Jupiter moving across the sky at about 15″ per second of time takes four or more seconds to disappear behind a remote tower, though diffraction prolongs the effect and makes accurate measurement impossible. In the sequel we shall assume that Venus at its nearest has a diameter of 1 arcminute.

It has been noted that the earth and the moon with radii 6360 km and 1700 km (figures soon to be verified) are completely different worlds because the retention of an atmosphere depends on surface gravity which is proportional to radius (though depending also on rock density). Thinking ahead to the planet Mercury, there seems to be a critical radius, around 3000 km, below which a planet near the sun is likely to be arid, without an atmosphere and of low albedo especially near half phase, while above 3000 km it can retain enough atmosphere to increase significantly its albedo.

The astronomical unit II

Just as Jupiter and Venus were compared, so Venus and Mercury were seen a few degrees apart after sunset on a brilliantly clear evening in Edinburgh on 19 March 1985. The positions of Venus being known as in Chapter 4, watch was kept on the separation of the two planets during this very favourable elongation. Mercury was estimated to be 18° from the sun on 19 March and close to greatest elongation and therefore with a phase near 0·5.

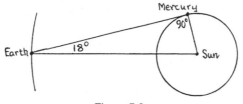

Figure 7.2

From Figure 7.2, where as usual the orbit is regarded as circular, $p=0·5$, $e=1\times\cos 18°=0·95$ AU, $s=1\times\sin 18°=0·31$ AU. The perspex plates produced a brightness ratio of 81:1, not easy to measure against the variable background twilight. Lastly, the Venus positions and phase were deduced from Chapter 4 results. Here are the values arranged as before.

Planet	Albedo a	Phase p	Solar distance s (AU)	Earth distance e (AU)	Brightness b
Venus	—	0·08	0·72	0·315	81
	(0·41)	(0·10)	(0·72)	(0·317)	(63)
Mercury	—	0·5	0·31	0·95	1
	(0·022)	(0·37)	(0·324)	(0·86)	(1)

(Albedos again modified for phase effects.)

First suppose that Mercury has a radius exceeding 3000 km and enough atmosphere to give it an increased albedo; we will try $a_V/a_M=3/1$. Then

$$\frac{r_1^2}{r_2^2} = \frac{1}{3} \times \frac{0·5}{0·08} \times \frac{0·72^2}{0·31^2} \times \frac{0·315^2}{0·95^2} \times \frac{81}{1}; \quad \frac{r_1}{r_2} = 10·0.$$

Do-it-yourself Astronomy

If this were so, Venus would have a radius of *at least* 30 000 km and Jupiter a radius of at least 400 000 km, with other consequences which are too improbable to be worth pursuing further.

Alternatively, suppose Mercury has a radius less than 3000 km and a barren rocky surface not unlike the moon. At half phase, with long shadows over an airless surface, the effective albedo will be near the lowest possible; let us try writing $a_V/a_M = 10/1$. This time the formula gives $r_1/r_2 = 5 \cdot 5/1$ from which Venus will have a radius *less* than $5 \cdot 5 \times 3000 = 16\,500$ km. But, being as we suppose a cloudy planet, its radius will also *exceed* 3000 km.

Therefore 3000 km<radius of Venus<16 500 km.

It being agreed that all this is not to be taken too seriously, and since the whole investigation is based on ratios, let us take as the radius of Venus the geometric mean of these tentative limits, this being the square root of their product, and equal to almost 7000 km. The actual value is 6052 km.

Finally if the diameter of Venus, regarded as 14 000 km, subtends at the earth an angle of 1 arcminute when the planet is at inferior conjunction and at a distance from us of 0·28 AU, then

$$0 \cdot 28 \text{ AU} = 14\,000 \cot(1') \text{ km and}$$

$$1 \text{ AU} = \frac{14\,000 \cot(1')}{0 \cdot 28} = 1 \cdot 7 \times 10^8 \text{ km},$$

which may perhaps on this one occasion appear alongside the true value, which is $1 \cdot 5 \times 10^8$ km, correct to two figures.

Notes on Chapter 7

Note 7.1. The neutrally tinted perspex sheet used for comparing brightnesses was of good optical quality and 2·5 mm thick; it was sawn up to provide a supply of similar square plates. The eye is better at comparing low levels of light; I arranged white cards at variable distances from a shielded small filament source in an otherwise dark room. When the distances were in ratio 3:1

84

and the intensities therefore in ratio 1:9, the brighter card seen through four plates appeared equal to the fainter card.

Thus the pile of four plates transmitted $\frac{1}{9}$ of the light falling on it, so a single plate would transmit $(1/9)^{0.25}=0.577$ of incident light.

In the Venus–Jupiter comparison three plates were needed; together they transmitted $0.577^3=0.192=\dfrac{1}{5.2}$ of the light.

In the Venus–Mercury comparison eight plates were needed; they transmitted $0.577^8=\dfrac{1}{81}$ of the light.

Note 7.2. *Venus–Mercury comparison.* If the more accurate though rounded values are used in the formula, $r_1/r_2=2.90/1$. In fact the ratio of the radii is $6052/2439=2.48/1$. Part of the discrepancy is again due to the compound effect of rounded decimals, but also the phase of Venus was small and changing rapidly, as was the effective albedo.

Note 7.3. *Artificial conjunctions of planets.* Very rarely there are conjunctions of planets so close that they appear together in the same field of a telescope, providing fascinating but all too brief direct comparisons of size, brightness and colour, factors with which we have been concerned. Though a little beyond the scope of this book, the following way of bringing together in a telescope the images of two planets which are well separated in the sky, may interest the owners of larger reflecting instruments of aperture eight inches (20 cm) upwards with well-aligned equatorial drives.

Figure 7.3 shows the arrangement though the mechanical support details are left to the imagination. Two extra flat mirrors are attached to the telescope on a light open frame across the end of the tube, in much the same way that four flats were fixed on a beam atop the 100 inch telescope at Mount Wilson to form A. A. Michelson's stellar interferometer. The extra flat X is set at 45° while the flat Y has some range of movement in two perpendicular directions. An aperture stop S defines two beams of light into the

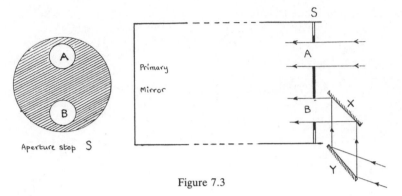

Figure 7.3

telescope, one from each object. Aperture B could be a little larger than aperture A to compensate for the light loss at the two extra reflections.

The first object would be sighted in the normal way using aperture A and kept in the field of view by driving the telescope. The second object would then be brought in by adjusting the mirror Y. This might prove to be a tricky operation but once done the two images should stay close together with only a slow relative drift provided the polar axis of the mounting was well aligned. If the sub-apertures A and B were 3 inches (8 cm) then a useful magnification could be employed. Back-silvered mirrors of reasonable quality would serve for X and Y at least for an initial trial, and the moon would be a useful second object, offset from the telescope axis, for finding the working position of mirror Y.

It might also be interesting to set a variable star such as Algol (β Persei) close to a star of comparable brightness such as α Persei, and look for the change in a few hours. If the mirror Y were left alone it would be easy to pick up the pair of stars in the field on the next occasion.

Note 7.4. *An occultation of Mercury by Venus.* Occasionally the actual eclipse or occultation of one planet by another has been recorded and it is interesting with the aid of a home computer to retrace and to some extent relive such an event. Here are three extracts from the *Philosophical Transactions of the Royal Society,*

The astronomical unit II

London (abridged), volume 8, pages 251 and 470, which refer to the occultation of Mercury by Venus as seen at Greenwich in 1737 by Dr John Bevis, well known in those days as an assiduous and careful observer. The date given is 17 May 1737 but this is in the Old Style Calendar and corresponds to 28 May (New Style). He wrote:

At 9^h 43^m 4^s p.m. app. time, Mercury was only the 10th or 12th part of Venus's diameter distant from her; afterwards the view was interrupted by clouds.

At 9^h 44^m, the planet Mercury appeared immediately under Venus, their distance asunder not more than a 10th part of the diameter of the latter.

At 9^h 51^m 10^s Venus shone out again very bright, and Mercury was quite covered by her. Afterwards clouds again prevented any further view of the phenomenon.

First we note that 'app. time' is apparent time as it would be given by a sundial at Greenwich, on that date about 2^m 50^s ahead of what we now call Greenwich Mean Time or Universal Time. So the times of observation were 28^d 21^h 40^m 14^s, 28^d 21^h 41^m 10^s and 28^d 21^h 48^m 20^s (UT).

With Dr Duffett-Smith's program, my computer gave the following details:

28 May 1737 (New Style)		Geocentric coordinates					
		Right ascension			Declination		
		h	m	s	°	′	″
21^h 40^m 14^s	Mercury	5	57	48·6	25	37	20·8
	Venus	5	57	50·6	25	38	1·6
				−2·0			−40·8
21^h 41^m 10^s	Mercury	5	57	48·8	25	37	20·8
	Venus	5	57	50·6	25	38	1·2
				−1·8			−40·4
21^h 48^m 20^s	Mercury	5	57	50·7	25	37	20·0
	Venus	5	57	50·2	25	37	58·0
				+0·5			−38·0

	Venus	Mercury
Elongation from sun (east)	22·16°	22·16°
Diameter (arcseconds)	52·3	6·9
Phase	0·07	0·55
Distance from earth (AU)	0·324	0·98

Sun's approximate position   RA $4^h 23^m$ Dec. 21° 35′ N

The latitude of Greenwich is   51·5°N

Using the spherical trigonometry formulae of Appendix 2 it may be shown that the planets were evening stars in the twilight sky to the north-west, 1 hour and 40 minutes after sunset, at an altitude of only 2° and descending towards the horizon at an angle of 30°. Seen through a telescope, Venus was a brilliant slender crescent alongside the paler smaller Mercury which showed a phase just greater than one-half. In those days Greenwich was well clear of London but it is remarkable that Dr Bevis managed to see as much detail as he did through an extreme thickness of turbulent air just above the outline of the city.

Concentrating on the more interesting second and third observations, Figure 7.4 has been drawn from the above differences of coordinates using the method given in the Notes to Chapter 1. A and B are the positions of Mercury relative to Venus as seen from the centre of a transparent earth and it seems clear that from such a viewpoint no occultation of Mercury was possible. But things are changed by the application of *parallax*, whereby the viewpoint moves from the centre of the earth to Greenwich along an earth radius which happened to be almost perpendicular to the direction to the planets (which were near the horizon).

To make the correction, hold the diagram out to the north-west with the plane of the paper vertical, then turn the page through 30° clockwise so the right ascension axis is at 30° to the north-west horizon, simulating the conditions of the original observations. Now imagine a rapid rise from the earth's centre up to the surface at Greenwich. Venus would drop past the remote stars by an amount equal to its own radius (being roughly the same size as the earth), while Mercury at three times the distance would drop only one-third as much. The net effect is that, relative to

The astronomical unit II

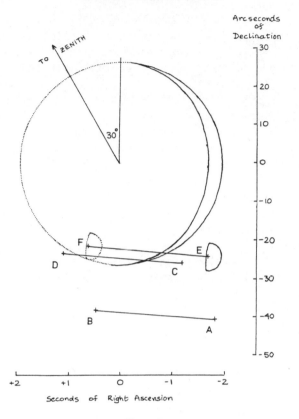

Figure 7.4

Venus in the diagram, Mercury would move *up* by two-thirds of the apparent radius of Venus giving the positions C and D which are now corrected for parallax.

Remembering that the crescent of Venus is so small that it is invisible to the unaided eye, the program and the computer have done well to take us thus far. Certainly the length AB or CD gives the magnitude of the shift during the 7^m 10^s between observations, the only doubt is in the location of the line of Mercury's progress. But as Mercury was first 'immediately under Venus' (remember that 30°) and subsequently 'quite covered', the most likely parallel track is from E to F, at which points the planet has been drawn to scale. We have the Doctor's word that the occultation was total but it was evidently a close run thing. For some further calculation details see the end of Appendix 2.

Note 7.5. Since writing the above I have read with interest the article by R. W. Sinnott and J. Meeus in *Sky and Telescope*, 1986 September, with extracts from the original *Philosophical Transactions*. Of the three quotations I have given from the abridged *Transactions*, the second now seems to be another version of the first but uncorrected for a clock error of 56^s. So the first and third form a compatible pair; the time interval between observations should be 8^m 6^s instead of 7^m 10^s and the lengths AB, CD and the displacement of Mercury on the Figure should be 13 per cent greater. Fortunately the general conclusion is as before, but I am glad to acknowledge the source of this correction.

8

The radius of the earth
✳ ✳ ✳

The first serious attempt to measure the circumference of the earth was made by Eratosthenes (275 B.C.–194 B.C.) who found that at midsummer noon the sun shone vertically at Syene on the Nile, while at Alexandria to the north its rays were inclined to the vertical at an angle equal to one-fiftieth of a revolution, or 7·2°. By measuring the distance between the two places and regarding it as one-fiftieth of the circumference, he obtained a good estimate of the size of the earth.

From the Polaris experiment in Chapter 2 we found it difficult without a telescope to measure absolutely an angle such as the altitude of a star to better than 2 arcminutes. So, for a project like this, any direct repetition of Eratosthenes' experiment is unlikely to give a useful result, even if an adequate baseline could somehow be measured across a hundred miles or more of intensively used countryside. We have seen that better accuracy is obtained whenever the steady rotation of the earth has enabled a succession of readings to be made and averaged; in particular the solar observations of Chapter 3 led to the detection of quite small changes in the sun's angular velocity. It seems therefore that to estimate the earth's radius from a reasonably short baseline, we have in some way to incorporate this technique of averaging when a small angular shift has to be found.

The following method was used in the summer of 1984 and involved two observing sites, one on a playing field near my home and the other some 27 miles away to the east, beyond the village of Spott near Dunbar. These formed a baseline shown as WE on Figure 8.1, where also EC is the required radius of the earth and EA is the radius of the small circle of constant latitude of which WE forms a part.

91

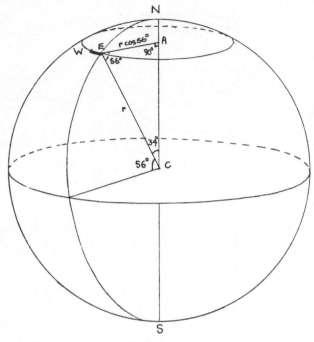

Figure 8.1

There are four facts which have been or will be established:

1. The latitude of the baseline is 56° 00' (Chapter 2).

2. The earth rotates relative to the stars in 86 164·1 seconds of mean solar time, just one sidereal day (Chapter 1).

3. The length of the baseline is 44·04 kilometres (this chapter).

4. The time for the earth to rotate through a small amount corresponding to WE is 169·8 seconds (this chapter).

With these results the radius EC=r km can be calculated. Because the earth rotates uniformly, WE is a fraction $\dfrac{169\cdot8}{86\,164\cdot1}$ of the circumference of the small circle.

Therefore $\dfrac{169\cdot8}{86\,164\cdot1}\times2\times\pi\times r\times\cos 56°=44\cdot04$,

The radius of the earth

$$r = \frac{86\,164 \cdot 1 \times 44 \cdot 04}{169 \cdot 8 \times 2 \times \pi \times \cos 56°}; \quad r = 6360 \cdot 5.$$

The conclusion is that the radius is 6360 km, correct to three figures.

It remains to show how the baseline was measured and how the time for the earth to rotate through WE was found. Both were enjoyable practical exercises during the best days of the summer, but I realise that although they were memorable sagas for me they are of limited general interest and I shall try to be reasonably brief.

The original steel foot ruler which I had resolved to make the basis of all distance estimates was used to measure accurately two steel pipes each some 2 metres long and with their ends filed flat. These pipes were placed end to end in contact on the ground and moved ahead alternately until a new reference length of some hundreds of metres was set up on a straight road, and probably correct to within a centimetre. To the rear wheel of my bicycle, well inflated, was attached a reliable revolution counter; the rear wheel being chosen because it follows a straighter course than the front steering wheel. The new reference length was then converted into wheel revolutions with various trial runs and also checked immediately before and after the main excursion.

There happens to be through Edinburgh and beyond a series of roads which trend in a generally eastwards direction; the whole journey to establish the baseline divided itself conveniently into 61 vectors joined end to end and each specified by a length and a compass bearing. To take the bearings I allowed myself the luxury of a small prismatic compass, remembering to stand well clear of the steel bicycle when using it. Starting from the west observing site very early one morning to avoid city traffic, the survey was soon under way rather like a round of golf on an enormous scale. I simply cycled in a careful straight line as far as possible, took a fresh bearing at the next bend and noted the revolution counter reading, and then on with the following vector. Figure 8.2 shows the route reconstructed from the 61 vectors; a map has only been used to add some well-known hill features in the area. Accuracy being all-important, I made rather leisurely progress in the increasing heat, but by the afternoon I reached the pleasant upland

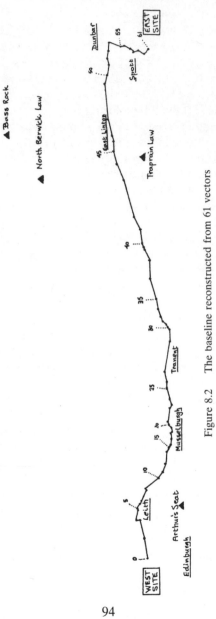

Figure 8.2   The baseline reconstructed from 61 vectors

The radius of the earth

country near the east site, with tremendous views west and north over the Firth of Forth. When the last vector was completed I cycled back to Dunbar and cooled off by the sea, then caught a train to Edinburgh and sped past the landmarks hard won earlier in the day. Some further comments on the survey are at the end of this chapter (Notes 8.1 and 8.2).

The value obtained for the west-to-east distance was 44 042 metres. The true value based on Ordnance Survey maps is 44 250 metres, from which the error is found to be 0·47 per cent.

The remaining problem is the timing of the earth as it rotates relative to the stars by an amount corresponding to the baseline WE; the time is in fact rather less than three minutes, as was mentioned earlier. This is equivalent to saying that a particular aspect of the celestial sphere will be seen about three minutes earlier at E than at W. To find this time difference precisely an observer must note the event or aspect at say the west site, travel next day to the east site and be ready for a recurrence of the event after about three minutes less than one sidereal day.

The aspects of the heavens which are more suitable for field experiments seem to involve either a local vertical, such as a hanging line stabilised by a weight in a bucket of water, or else a local horizontal surface as defined by a vessel containing light grade oil. The first aspect I tried to capture was the moment when the bright stars Bellatrix and Rigel in the constellation Orion appeared in vertical alignment; at 56° N this happens a little over one hour after Orion is in the south. Figure 8.3 shows Orion (a) when due south, (b) at the moment of alignment, and (c) some time later. Anyone recalling the introduction to this book will now understand why I was by the Witches' Stone on that evening in March. I will only add that the attempt to time the transits of the stars past a vertical line seen through a series of viewing holes, although interesting in theory, was almost useless in practice and for two reasons: the wind disturbed the vertical line and the method was insensitive, it taking about 7 seconds of real time for the interval between transits of Bellatrix and Rigel to change by 1 second.

Up to this time I had been thinking of various aspects of the celestial sphere of stars, necessarily involving observations by

95

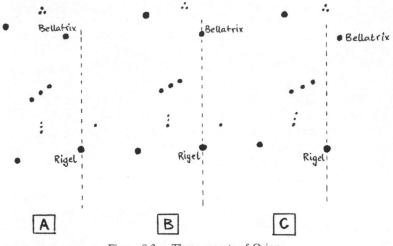

Figure 8.3   Three aspects of Orion

night. To observe the sun with its rapid and variable motion along the ecliptic seemed hopeless. Eventually it dawned that although the motion of the sun is complicated, it is changing in a continuous manner; movement along the baseline would show up as a discontinuity in a chain of observations just as surely as a scratch on a record shows up against a complex but continuous background of sound.

For reference purposes I would use not a local vertical but a local horizontal in the form of a dish of thin engine oil which would give an accurately horizontal reflecting surface untroubled by slight tremors. A narrow beam of sunlight could be reflected in the oil surface; moreover it could be shut up in a box to defeat the wind problem. The aspect to be caught on a succession of days would be the moment when the sun, low in the west in summer, reached a definite altitude around 20° depending on the observing device. The daily timing of this event would change with the advancing season in a continuous manner but showing a discontinuity when the observation switched from one end of the line to the other. When I realised all this there was just enough time, given good weather, to try it before the summer was over.

The radius of the earth

Incidentally, the westerly position of the sun was important. It can be shown that any error due to a slight north-to-south difference between the ends of the baseline is then insignificant (Note 8.3).

The final device which became known as the Bridge is shown in Figure 8.4. Made as light and rigid as possible, it was supported only at its ends, the locating lugs LL dropping onto support brackets fixed to the containing box which is shown as a broken line. In this way the mechanical strain on it was minimised and it was protected from the elements.

H is a 3 mm hole which defines a pencil of sunlight, D is a dish exactly positioned and containing a metered amount of oil acting as a mirror; S is a viewing screen of millimetre ruled tracing paper fixed to the underside of a sheet of glass, the underside avoiding trouble from internal reflections. P and T are a glass plate and a tray of oil; the reflections from these were used to level the Bridge and generally every precaution was taken to make conditions the same for each daily run of observations.

Light entered the front of the box through a 4 cm hole and was restricted to a narrow beam by the inside hole H. As the sun in the west slowly declined, the beam from H advanced across the oil surface like a spotlight moving across a stage; the reflected ray rose and met the screen where it appeared as a small elliptical patch when viewed from above through a slot in the box lid. During each daily experiment the image moved down the same long diagonal of the rectangular screen; as it passed each millimetre mark

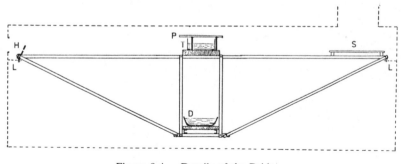

Figure 8.4   Details of the Bridge

it was timed with clock and tape recorder as in Chapter 3, and from all this a best estimate was made of the moment when reflection was from the middle of the oil tray (and known by geometry to correspond to a certain screen reading). This was the sought-after event.

The optical arrangement has the important property that if the fore and aft levelling of the Bridge is in error by a few arc-minutes, the position of the light spot on the screen is practically unaffected – it is as if a servo-stabiliser were acting, provided reflection takes place from near the middle of the oil tray. This means that the averaging of many readings taken during a run will genuinely improve accuracy, and is not just a pretence. Figure 8.5 shows the degree of correlation that was obtained with the device.

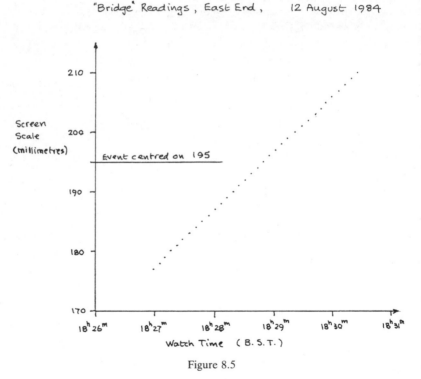

Figure 8.5

98

The radius of the earth

A settled spell of sunny weather led to the following event times:

1984 August 10	west	$17^h\ 35^m\ 28^s{\cdot}9$	
August 11	west	$17^h\ 33^m\ 32^s{\cdot}5$	
August 12	east		$17^h\ 28^m\ 46^s{\cdot}4$
August 14	west	$17^h\ 27^m\ 43^s{\cdot}6$	

These are shown in Figure 8.6 to which has been added the time at the west site for 12 August obtained by contructing a best-fitting quadratic curve (almost a straight line) through the three west points.

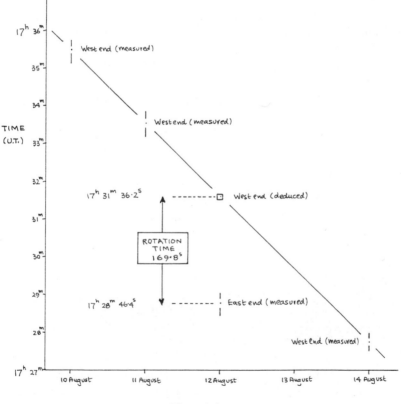

Figure 8.6

99

The rotation time corresponding to the base-line WE is found by subtracting the measured time at the east site, 12 August $17^h\ 28^m\ 46^s\cdot4$
from the deduced time at the west site, 12 August $17^h\ 31^m\ 36^s\cdot2$

giving the result $169^s\cdot8$

This is the fourth and final item required for the calculation at the start of this chapter.

During my exploration of the pleasant hill country around the east site I discovered in the village of Innerwick a remarkable sundial. I realised that it must be just about east of my home in Edinburgh, and then discovered from the Ordnance Survey that it is absolutely due east, to the very yard or metre, though some 30 miles away. It dates from the year 1893 and on it are these rather wistful lines:

> O'er every hour that's brightest
>   A shadow creeps,
> And he whose laugh is lightest
>   Full often weeps.
> Oh look we for the morrow
>   Which hath no night,
> Where lost is every sorrow
>   In God's own light.

Notes on Chapter 8

Note 8.1. *The chain of vectors.* If the vectors are given in the polar form (metres, bearing), for example (1320, 078°) and (1570, 095°), then the west-to-east distance is the sum of 61 terms of the type

$$- - - + 1320 \sin 078° + 1570 \sin 095° + - - -.$$

On a gentle bend it can be difficult to judge the end of one vector and the start of the next, but provided they trend generally eastwards this is not too serious. For example, the above two vectors would give a contribution 1320 sin 078°+1570 sin 095°=2855·18 metres, but if the end point is moved 20 metres, shortening one and lengthening the other, the contribution of the pair is now

$$1340 \sin 078° + 1550 \sin 095° = 2854·82 \text{ metres,}$$

which is a difference of only 36 centimetres.

Note 8.2. *The north–south error.* The north-to-south difference between the sites is likewise the sum of 61 terms in which cosine replaces sine. Unlike the east-to-west series, this one is liable to considerable cumulative error; my survey put the east site 390 metres south of the west site but the actual distance is 1500 metres. As mentioned in the main text this is of no consequence if the sun is observed in the west. For suppose that an observer could be rapidly transported a few hundred kilometres back and forth along a meridian in a few seconds. The celestial sphere would appear to move forth and back about the line joining the east and west horizon points; an object in the west would move at the top of a short circular arc and its altitude would hardly change.

Note 8.3. *The 'servo-compensation' effect.* Calculation shows that if the Bridge were tilted by 0·2° which is easily detected by the optical levelling device, the image on the screen is disturbed by about one-seventh of a millimetre; this would shift the mean time derived from a run of readings by a little under a second.

9

The moon
✳ ✳ ✳

There is such endless interest in exploring the moon with a telescope that anything said here must appear minor by comparison, but to justify statements made in other chapters we should at least verify the moon's approximate distance and size. This can be done by geometry based on a study of the earth's shadow at a lunar eclipse, assuming the sun to be very distant by comparison, and such arguments have been used from very ancient times. The principal shadow or umbra is always found to have a circular boundary, and it must have been disconcerting for the early astronomers to discover that the earth, judging by its shadow, is not only round but apparently unsupported. The diameter of the earth's shadow, when deduced from the part that is seen lying across the moon, is invariably about three times the diameter of the moon itself, so the whole shadow would subtend an angle of 1·5° while of course the moon and sun subtend angles of 0·5°.

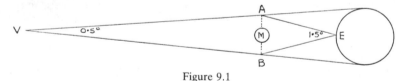

Figure 9.1

In the Figure, V is the vertex of the conical shadow of the earth. Seen from V, the earth would just eclipse the sun and so angle AVB=0·5°. But as mentioned above, angle AEB=1·5°, therefore ME is roughly one-third of VM or one-quarter of VE. But VE is 240 times the earth's radius by the one in sixty rule, so ME is about 60 times the earth's radius, say 380 000 km or 240 000 miles.

As the moon subtends half a degree, the one in sixty rule

102

again tells us that the diameter of the moon is found by dividing by 120; the result is 3170 kilometres or 2000 miles.

A completely different approach to the problem is to use Kepler's third law to link the motions of an earth-skimming satellite with the motion of the moon. It was mentioned in Note 6.3 that the third law is involved with the inverse square law of gravitation which holds good for external points right up to the earth's surface. This is because of the important result, discussed in Appendix 1, that the earth attracts external objects as if it were a single massive particle replacing the real earth and located at its centre.

In 1666 Newton attempted to use the inverse square law to make this link between the moon's motion and gravity at the earth's surface but was set back through using a value for the earth's radius which was later shown to be incorrect, and he put the matter aside for thirteen years. On hearing through Robert Hooke of a new determination of the radius by J. Picard, he repeated his calculation in 1679 and found a satisfactory fit. But a further six years elapsed before he proved the result in Appendix 1 and which finally set all the details in order.[7]

The formula to be used is Kepler's law in the form

$$\frac{r_1^3}{r_2^3} = \frac{t_1^2}{t_2^2} \text{ in which}$$

r_1, t_1 are the moon's distance and sidereal period of rotation, and r_2, t_2 are the orbit radius and sidereal period for the satellite.

The quantity t_1 is the sidereal month and can be found by noting the moon's progress past bright stars, but as time goes on one discovers that the motion is far from simple and the sidereal period can vary by some hours. This was a major problem for the old navigators trying to use the moon as a celestial clock that would enable them to find longitude at sea, but for our calculation an approximate figure is sufficient. For example I observed the moon in the neighbourhood of the stars Castor and Pollux on 22 December 1983 and again after eleven revolutions near the same stars on 18 October 1984, obtaining a value for the sidereal month of $27^d 8 \cdot 6^h$, nearly an hour longer than the recognised long-term mean period, which is an error of one-sixth of 1 per cent.

Do-it-yourself Astronomy

Also needed is the acceleration due to gravity near the earth's surface. This I found by timing 100 small swings of a pendulum consisting of a golf ball suspended by a fine thread, the length to the centre of the ball being 2·450 metres and the total time being 314·0 seconds on various occasions. The standard formula $t=2\pi\sqrt{(l/g)}$ gave $g=9·81$ metres per second per second. We are now ready for the moon distance calculation.

First remove the atmosphere and launch the skimming satellite:

$$r_2=6360\,\text{km}=6·36\times10^6\,\text{m} \qquad\qquad \text{(Chapter 8)};$$

$$g=9·81\,\text{ms}^{-2};$$
$$v^2=rg; \quad v=7899\,\text{ms}^{-1};$$

$$t_2=\frac{2\pi r}{v}; \quad t_2=5059\,\text{s}.$$

For the moon,

$$t_1=27^{\text{d}}\,8·6^{\text{h}}=2\,363\,760\,\text{s}$$

$$r_1{}^3=r_2{}^3\times\frac{t_1{}^2}{t_2{}^2}; \; r_1=r_2\times\sqrt[3]{\frac{t_1{}^2}{t_2{}^2}} = 382\,957\,\text{km}$$

The estimate of the moon's distance is 383 000 km.

The moon is never very far from the celestial equator and is seen to move past a distant object in two minutes, from which its angular diameter is close to 30 arcminutes and consequently its radius in kilometres is 383 000 tan 15′ which is 1670 km. Because the angle estimate is only approximate this figure is 4 per cent too low, the true radius being 1738 km, but either way the moon is seen to be about one-quarter the linear size of the earth. Though it is far from being a tiny companion, the reduced surface gravity has made it incapable of retaining an atmosphere and the lack of erosion has preserved every feature of its history. This extreme difference between the two worlds was crucial in the argument used in Chapter 7.

There is an interesting but sometimes puzzling phenomenon concerning the moon's appearance in the sky. When the sun and moon are both visible but well separated, for example when the moon is quite high in the south-east and the sun is low in the west, the moon's sunlit hemisphere is illuminated from well above the horizontal; at the same time the sun which is the source of the light is lower in the sky than the moon as in Figure 9.2. At first

Figure 9.2   The moon paradox

105

glance it seems quite impossible. But take a string and hold it out tight between outstretched arms so it passes high in front of your head and joins moon to sun with what is undoubtedly a straight line. It rises from your left hand and descends to your right hand: the brain is trying to interpret the extended panorama and is deceived! The straight string and the eye together define a plane which if produced will intersect the celestial sphere in a great circle arc joining the sun and moon by the shortest route *on the sphere*. A similar effect arises with great circle journeys between well-separated places on earth. The shortest route from the British Isles to Vancouver surprisingly heads out north-west towards the Arctic before turning back towards the south-west. Stretch a string between the two places on a terrestrial globe and it all makes sense.

Both situations involve great circle arcs joining two places in general position on a sphere and such problems lead naturally to the study of spherical trigonometry. Much of interest can be tackled with one key formula which is not hard to prove if the scalar product of two vectors is understood. Appendix 2 gives an introduction to this topic and some illustrations of its use in navigation and astronomy.

10

The nearer stars

✳ ✳ ✳

If it is reasonable to regard our sun as a fairly typical star in order to avoid giving it a specially privileged place in the universe, and if the distribution of stars in our neighbourhood is nearly uniform, then by comparing the brightness of the sun with that of a nearby star one can form some idea of the enormous distances separating the stars: about a quarter of a million astronomical units. If this letter O represents the earth's orbit round the sun with a diameter of 2 AU then the nearest stars would be a quarter of a million millimetres away, about 250 yards or metres.

Two examples of this kind of estimating will be given: onc is modern and the other is of some historical interest and shows the early approach which at least prepared scientists to expect very large answers. For if the figure of a quarter of a million AU is near the mark, it follows that when the earth is displaced across its orbit by 2 AU in six months, the angular shift of a nearby star against the remote stellar background is only arctan (2/250 000) which is 1·65 arcseconds. Two or three centuries ago no one was sure that the sun was not unusually dim or bright and it was important to confirm the above rough estimate by direct measurement of the small angular shift, but the practical difficulties were such that success was delayed until the year 1838.

The first of the comparison methods may seem rather naïve but was undertaken as the final step in the sequence of measurements linking my steel foot ruler with the stars. The object was to compare the brightnesses of the sun and the northern star Capella using Venus as an intermediate standard. The equipment was simple enough, consisting of a car wing mirror and a spherical reflecting glass mirror ball. To begin with, one has to understand how the upright diminished image typical of a wing mirror or other convex reflector is produced.

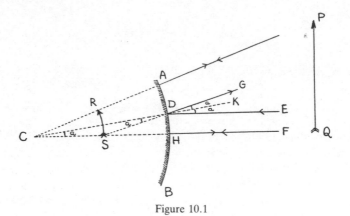

Figure 10.1

In Figure 10.1, AB is the convex spherical reflecting surface with centre of curvature at C. The arrow PQ is to suggest an object which is really a large distance to the right compared with ABC, and whose image is to be constructed. The lines CA, CDK, CSH are radii and the four angles marked *a* are equal by geometry and the law of reflection.

The first ray FH from the direction of Q strikes the mirror at H and is reflected back along its own path, as if from S. The second parallel ray ED from the direction of Q strikes the mirror at D and is reflected along DG, as if from S. Therefore the (virtual) image of Q is at S and, if the angles are sufficiently thin, CS=SD=SH, so S is halfway along the radius CH. Likewise the image of remote P is at R, halfway along the radius CA and the reduced, virtual image of PQ is RS as shown. Thus the picture surface for remote objects is located midway between the mirror and its centre of curvature. For a car mirror the radius of curvature is about 4 feet and the location of the picture surface can be checked by moving one's head slightly from side to side.

If the remote object PQ subtends an angle θ at the eye when seen direct, so that angle PCQ=θ, then also angle RCS=θ and angle RHS=θ. Moreover the image when viewed from in front of the mirror at a distance from S that is *n* times SH (that is, *n* times half the radius of curvature), will subtend at the eye an angle of

108

θ/n. When driving, n is about 2 and the vehicle behind is seen with its linear dimensions halved. By retreating from the mirror so that n increases, the angular size of the image soon becomes tiny. What we now have is a way of reducing the apparent size and brightness of a remote source by a factor which can be stated with fair precision. In addition we need to remember that the area of the source is reduced n^2 times, and so also is the brightness.

Just before dawn on 3 January 1987 Venus was fairly low and a little east of south seen from Edinburgh, while Capella was directly opposite, somewhat west of north and at much the same altitude. The wing mirror of my car has a semi-radius of curvature of 70 centimetres; it was adjusted until I could see both Capella direct and Venus reflected, as close together as possible. When 9 metres away I judged the brightnesses to be equal. The value of n was therefore 970/70 from which $n^2 = 190$ and Venus is found to be 190 times as bright as Capella. (For some discussion of these results see Note 10.1.)

A few hours later the sun had replaced Venus and the morning was crisp and blue. I suspended the glass mirror ball, radius 25 mm, which had just seen service as a Christmas tree decoration, so that seen from the south it shone against the sky background. I had often seen Venus in daylight and my problem was to judge the point from which the little sphere appeared as bright as Venus would have been if in the northern sky.

After much pacing around I settled on a distance of 150 metres from the sphere. This gave $n = 150\,000/12\cdot5 = 12\,000$ and $n^2 = 1\cdot44 \times 10^8$ and the sun is found to be $1\cdot44 \times 10^8$ times as bright as Venus. Combining this with the previous result we find the sun to be $190 \times 1\cdot44 \times 10^8$ times as bright as Capella, and on taking the square root and applying the inverse square law, the distance of Capella (assumed to be similar to the sun) is 165\,000 times the distance of the sun, 165\,000 AU. This is about two and a half light years, only half the true mean separation of stars in our neighbourhood. As there were errors which tended to cancel one another I must consider myself lucky with the above result (see again Note 10.1).

The second method of comparing brightness is due to Sir Isaac

Newton and is outlined among a few manuscript sheets which were no more than a rough draft.[8] Written perhaps 80 years after the invention of the telescope, it assumes the mean radius of Saturn (without its ring) to be 10 arcseconds, which is 1·5″ too large, and that Saturn when the rings are invisible appears like a star of the first magnitude. He refers to a 'hemisphere of Saturn's orb' by which is meant a hemispherical surface with its centre at the sun and a radius equal to the radius of the orbit of Saturn. There is no direct mention of this radius but it would be known to be about 10 AU. He is speaking of ratios, in the style 'A is to B as C is to D':

> The whole light wch comes from a hemisphere of the Sun or falls upon a hemisphere of Saturn's orb is to so much of his light as falls upon the body of Saturn as the surface of the hemisphere of Saturn's Orb is to the surface of Saturn's apparent diameter [i.e., disc], that is as the Radius of a circle to the versed sine of 10″, or as 10 000 000 000 to 11¾ or about 900 000 000 to 1.

In Figure 10.2 let the circle have radius 1 unit, then θ is the arc AD, the sine of θ is AB, the cosine of θ is OB and the

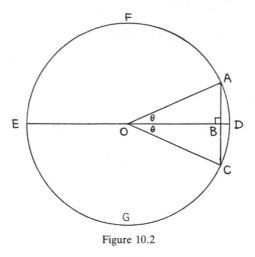

Figure 10.2

versed sine or versine or 'turned sine' of θ is BD; it is equal to $(1-\cos \theta)$.

By the intersecting chords result for a circle, $EB \times BD = AB \times BC$; $(2-BD) \times BD = AB^2$ and if θ is small the bracket tends to 2 so that $2BD = AB^2$; 2 versine $\theta = AB^2$.

Now regard the figure as a cross-section of the sphere of Saturn's orb so the sun is at O and AB becomes a radius of Saturn's disc (literally a disc). Angle θ is now $10''$ and

$$\frac{\text{area of hemisphere FDG}}{\text{area of disc AC}} = \frac{2 \times \pi \times 1^2}{\pi \times AB^2}$$

$$= \frac{2 \times \pi}{\pi \times 2 \times \text{versine } 10''} = \frac{1}{\text{versed sine } 10''}$$

which is Newton's result.

Next, $AB = AD = \theta$ very nearly, so versine $\theta = \theta^2/2$. Since $10'' = \dfrac{10}{3600} \times \dfrac{\pi}{180} = 4\cdot85 \times 10^{-5}$ radians, versine $10'' = 1\cdot175 \times 10^{-9}$ as given by Newton, and $\dfrac{1}{\text{versine } 10''} = 9 \times 10^8$ nearly.

The conclusion reached in this first quotation is that half the light from the sun, that is the half concerned in illuminating Saturn, is 9×10^8 times the sunlight intercepted by a hemisphere of Saturn (seen as a disc) and reflected back to us on earth, assuming perfect reflection. He continues:

And since in removing from a lucid body the light thereof decreases in a duplicate proportion of the distance, if we were 30 000 times remoter from the sun than we are at present, [he means 'remoter from the sun than we are from Saturn at present'] his light [the sun's light] would appear equal to that of Saturn without his ring or to that of a star of the first magnitude.

This means that the light decreases inversely as the square of the distance, and 30 000 is just the square root of 9×10^8. It is helpful to think of the earth and Saturn as fixed, but the sun removed to 30 000 times its present distance *from Saturn*. If Saturn were somehow illuminated as before, the sun and Saturn would appear equally bright seen from the earth. In addition we need to know that the mean distance of Saturn from us is about 10 AU, then the distance of a star of the first magnitude and similar to the sun works out to 300 000 AU, about $4\frac{1}{2}$ light years.

In Section 57 of *The System of the World* which is appended to the third edition of *Principia*, Newton briefly summarises this whole approach having corrected the planet's radius and given it an albedo of one-quarter. He also supposes Saturn to be rather brighter than a first magnitude star. After these changes the above distance is raised from 300 000 AU towards one million AU, about 15 light years.

Note on Chapter 10

Note 10.1. *The first experiment reviewed.* The two parts were each rather seriously in error, though these errors cancelled to some extent in the final result. Before considering the details, here is a brief mention of the system of stellar magnitudes which form a logarithmic brightness scale. Fainter objects have more positive magnitudes and a difference of 5 magnitudes, say from 1 to 6, or 0·3 to 5·3, or $-3·2$ to 1·8, or -12 to -7, corresponds to a brightness ratio of 100 times. For example, Sirius has magnitude $-1·46$ and Bellatrix (much fainter) has magnitude 1·64, the difference being 3·1. On a calculator take the fifth root of 100 which is 2·511 886 4 and raise it to power 3·1 obtaining 17·378; thus Sirius is 17·4 times as bright as Bellatrix.

We may now check the experiment using magnitudes

$$-26·8 \text{ (sun)}, \quad -4·6 \text{ (Venus)}, \quad 0·08 \text{ (Capella)}.$$

As Venus and Capella differ by 4·68 magnitudes, Venus was on that date 74·5 times as bright as Capella, from which the correct value of n was $\sqrt{74·5}=8·63$ and the distance from the wing mirror image should have been 8·63×70 cm, almost 6 metres instead of 9·7 metres, an error difficult to explain.

Again, the magnitudes of the sun and Venus differ by 22·2, so the sun was 7·58×10^8 times as bright as Venus, not 1·44×10^8 times, and I should have been 344 metres from the sphere, not

150 metres. The small image of the sun in the sphere has a diameter of only 12·5 tan 0·5° millimetres, about one-tenth of a millimetre, and the image seen from far off is formed by a tiny patch of the mirror surface measuring a tenth of a millimetre each way. On close inspection afterwards there were imperfections in the sphere's surface of this order of size, which no doubt affected the result.

Lastly, as the magnitudes of the sun and Capella differ by 26·88, the ratio of their brightnesses is $5·65 \times 10^{10}$. The square root of this is 237 684 from which the suggested mean separation of the stars near us is about 240 000 AU, or 3·8 light years.

11

The remote background
✳ ✳ ✳

Having reached out however tentatively and inaccurately to a distance of several light years from the earth, it might be thought that the opportunities of using improvised equipment would finally have run out. And yet there was one more experiment, first seen long ago in a science museum, which seemed especially worth trying at first hand: the mysterious pendulum of Jean Bernard Léon Foucault which he demonstrated in Paris in 1851. Apparently so simple, it is one of the most remarkable experiments in physics and its eye-opening potential should earn it a place in any general science course.

As a preliminary, consider a frame placed on a movable turntable as in Figure 11.1, carrying a weight on the end of a thread and with the point of suspension set exactly above the vertical axis of rotation of the turntable. The weight is given a slight push to set it moving as a pendulum in any chosen vertical plane and the turntable is then gently moved around. The result is as expected: the pendulum continues to oscillate in the same plane relative to the room, the stationary point of suspension being quite unable to pass on the frame's rotation through the fine thread to the pendulum below.

Foucault's idea was to show directly in this way the rotation of the earth. Before discussing the practical details, let us travel in imagination to the south pole (a better building site than the north pole) where a free pendulum, long and heavy enough to oscillate for six hours or more, has been suspended from an almost frictionless support in a massive weatherproof building. It is a rare cloudless day in early October and the sun is moving round the horizon at a practically constant altitude of 4°, having recently made its welcome return after the long months of winter darkness,

114

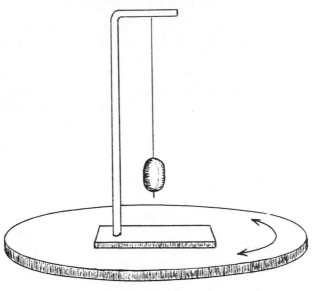

Figure 11.1

and its light is streaming into the pendulum building through the windows thoughtfully provided. The pendulum is started up exactly in line with the sun's centre and quite soon its direction of swing is indeed seen to be following the sun round the horizon, thereby showing that it is the building that is moving, just like the turntable. But wait! After six hours an extraordinary thing has happened: *the pendulum is gaining on the sun* and is ahead by almost a solar radius, a quarter of a degree in azimuth. Eventually there can be no doubt that the pendulum is ignoring the sun which moves along the ecliptic at a rate of one degree per day, and is following the remote stellar background. . . .

In reality the experiment was first done in the more practical surroundings of the Panthéon in Paris, using a pendulum 220 feet long (67 m) and with a bob of mass 62 pounds (28 kg). At either pole the rotation rate is 15° per hour but at latitude ϕ it is only 15 sin ϕ degrees per hour. This means that at Paris (49°N), New York (41°N) and Edinburgh (56°N) for example, the respective rates are 11·3, 9·8 and 12·4 degrees per hour. The object of using a long heavy pendulum is to delay the inevitable damping of the oscillations by air drag, yet with the most elaborate arrangement it would be difficult to sustain the motion for six hours with the necessary accuracy to confirm the sidereal rate rather than the solar rate. But as the pendulum rate is always found to be constant *regardless of the position of the sun or moon*, we need little convincing that these bodies have almost nothing to do with the phenomenon.

It is well worth while trying to see the rotation at first hand. While the most convincing demonstrations are found in lofty museum halls, something of the effect can be seen in home surroundings if a vertical height of four metres or more is available, reasonably free from draughts. In my own arrangement the suspension was fixed to a main roof beam from which the line descended through the roof space and the room below via an open trap door, giving a length of 13 feet (4 m). To the line was attached a longcase clock weight of 12 pounds (about 5 kg) with a vertical pointer fixed under it, just above a sheet of graph paper. The best clock weights are enclosed in polished cylindrical brass cases which would help to reduce air drag. The pendulum must be given plenty of time to stretch, unwind itself and completely settle down. The top suspension must be strong and immovable otherwise the energy of the bob will bleed away through it comparatively quickly. If the pendulum is started with too large an amplitude or if the suspension is at all sloppy, the motion will soon degenerate into an ellipse which masks the true effect. I found it best with a 4 metre length to start with a movement of no more than 8 centimetres either side of the vertical; this continues with little loss for an hour by which time the rotation is obvious and can be measured with the graph paper. Inevitably the path will assume the form of a long thin ellipse due to small unsymmetrical drag forces, but the long axis of the ellipse will still indicate the pendulum direction with sufficient accuracy. Great care must be taken to set it off to a smooth start, for example by arranging a block of wood to press on the weight at the level of its centre of gravity; the weight is pushed aside by the block, allowed plenty of time to settle to rest, and the block withdrawn positively but not jerkily so the resulting motion is exactly straight.

Although the rotation is to be expected, I still find it uncanny that, located in this solid city on a grey day when the weather is doing its worst, the pendulum will try to follow the celestial background *with not a star in sight*. How much more remarkable had we lived in a part of the galaxy surrounded by enough interstellar dust to limit visibility to say 200 000 AU, effectively hiding all the stars from view but with no obvious change in the appearance of the sun or planets. To explain the motions of the solar system in a sky devoid of stars would have been a long slow process for the early scientists; Kepler's third law and the idea that the earth rotates round the sun would point to a special kind of reference frame (set of coordinate axes) somehow preferred by nature. It would be seen that centrifugal forces tended to zero as local mechanical systems became non-rotating in such a frame, also that the mysterious pendulum behaved with an unerring sense of such frames of reference. But why was there this constraint, apparently related to nothing external? An alert philosopher might have argued that there must be some remote, invisible and substantial part of the universe that was influencing local events in the solar system. In due course radio astronomy would be invented, penetrating the cosmic fog and allowing dramatic access to facts about the distant scene. It would be realised that the special frames were simply non-rotating and non-accelerating with respect to the bulk of the material in the universe, and it would be natural to conclude that this remote material was directly affecting local events.

We who have no difficulty in seeing some thousands of stars even without a telescope, still find it hard to believe that they are affecting every detail of mechanical behaviour here on earth, from Foucault's pendulum to the domestic spin drier. For if the nature of the coupling is such that our dominant sun makes a negligible contribution, can we expect much more from the other stars we can see? Is their combined mass and influence sufficient to account for the highly isotropic (same in all directions) effects associated with centrifugal force? Spin driers do not perform noticeably better at special times in the sidereal day when their axes of rotation are

117

pointing to certain constellations! In the end, and within the equipment limits we have set ourselves, we are not much further on than the dwellers in the cosmic fog who had to postulate that a major part of the universe lay for ever hidden from their sight – this seems to be true for us as well.

Just why this remote matter has a low visibility would take us beyond the scope of this book and has to do with the well-established expansion of the universe. But the sheer extent of this far off but significant material was suspected even in the eighteenth century. For as we look out on the Milky Way we do seem to belong to a disc-shaped system, flattened perhaps by its own rotation. If so, then that rotation is relative to some vaster massive background within which we are embedded and by which we are influenced – only a universe on this scale can account for the isotropic mechanical features we find here on earth. Moreover, according to some cosmologists, just as every droplet in a shower mirrors the external world in which it is set, so every local gravitational force, such as the pull of the earth on a packet of sugar, depends for its magnitude on this remote influence. If the universe were not as big as it is, then even our local scene would be unthinkably different.

It would appear then that we are surrounded by an unseen cloud of creation which through the agency of gravity aligns the details of our lives. Perhaps Paul, who wrote something similar in his letter to the Hebrews, would not have been surprised to find that the physical world is constructed in this way.

Appendix 1. The attraction due to a sphere
✳ ✳ ✳

In Chapters 6 and 9 mention was made of the important and remarkable fact that a sphere attracts external masses as if it were a single particle at its own centre and with the same total mass. The sphere must have symmetry, that is, it must consist of spherical layers or shells each with a uniform density, though this density may vary with radius, as with a planet having a dense core and an extensive but more rarified atmosphere.

The result was suspected by Newton for many years before he finally proved it in 1685. Together with an associated result about internal points it appears in *Principia* Book 1 as Propositions 70 and 71. These were drafted in more modern form by W. Thomson and P. G. Tait in their *Treatise of Natural Philosophy*, 1867, Article 471 and this was further simplified by A. S. Ramsey, *An Introduction to the Theory of Newtonian Attraction*, 4th edn 1952 (Cambridge University Press) on which this account is based.

To begin with, here is a note on solid angles.

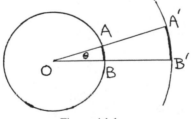

Figure A1.1

Given two concentric circular arcs as shown, the ratios $\dfrac{\text{arc AB}}{\text{OA}}$ and $\dfrac{\text{arc A'B'}}{\text{OA'}}$ are equal by similar figures, each being a measure in radians of the angle θ at the centre. As the circumference formula is $2\pi r$, the entire central angle is $2\pi r/r = 2\pi$ radians.

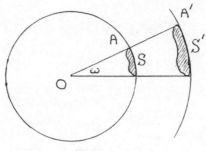

Figure A1.2

In the same way, given two concentric spheres and a cone vertex O meeting the spheres in surfaces S and S', then the ratios $\dfrac{\text{surface S}}{\text{OA}^2}$ and $\dfrac{\text{surface S}'}{\text{OA}'^2}$ are equal by similar figures, each being a measure of the solid angle at the centre. As the surface of a sphere is given by $4\pi r^2$, the entire solid angle at the point is

$$4\pi r^2/r^2 = 4\pi.$$

Theorem 1. The total attraction due to a uniform spherical shell of matter is zero at any internal point.

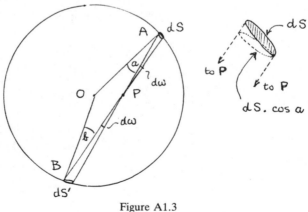

Figure A1.3

120

Appendix 1. The attraction due to a sphere

Let P be the internal point of the sphere centre O and let a thin double cone of solid angle $d\omega$ meet the shell in elements of area dS at A and dS' at B. The element dS is perpendicular to OA and its projection perpendicular to PA is $dS.\cos a$, where angle OAP=a. Also $d\omega=dS.\cos a/PA^2$.

If the surface density of the shell material is m, the attraction at P due to the small mass $m\, dS$ at A is along PA and proportional to $m\, dS/PA^2$ by the inverse square law, and hence proportional to $m\, d\omega/\cos a$.

Likewise the attraction at P due to the mass $m\, dS'$ at B is along PB and proportional to $m\, d\omega/\cos b$ where angle OPB=b. But $a=b$ because OA=OB. Therefore the net attraction at P due to this pair of elements is zero. Likewise for all such pairs of elements into which the shell can be divided by constructing double cones through P, and the result follows.

Theorem 2. The total attraction due to a uniform spherical shell of matter at an external point is the same as that due to a particle having the shell's mass and located at its centre.

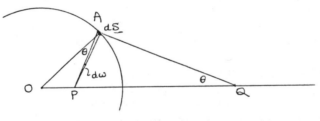

Figure A1.4

Let Q be the external point, then P is chosen to be the inverse of Q with respect to the circle, that is, OP×OQ=OA^2 or OA/OP=OQ/OA. Then triangles AOP, QOA are similar, having angle O in common and pairs of sides about O in proportion. From the similarity, angle AQO=angle PAO=θ; also OQ/QA=OA/AP. If m, dS and $d\omega$ are as before, the attraction at Q due to the

121

small mass $m \, . \, dS$ at A is along QA and proportional to $m \, dS/QA^2$ by the inverse square law, and the resolved part along QO is

$$\frac{m \, dS \cos \theta}{QA^2} = \frac{m \, dS \cos \theta}{AP^2 \times OQ^2} \, . \, OA^2 = \frac{m \, d\omega \, . \, OA^2}{OQ^2},$$

$$\text{because } d\omega = \frac{dS \cos \theta}{AP^2} \text{ as before.}$$

Summing for the whole shell, the total force along QO is $\dfrac{m \, . \, 4\pi \, . \, OA^2}{OQ^2}$ because the total solid angle subtended by the shell at the internal point P is 4π. As the shell area is $4\pi \, . \, OA^2$, its mass M is $m \, . \, 4\pi \, . \, OA^2$ and so the force along QO is M/QO^2 as required.

Lastly, if a solid sphere has symmetry, being made up of shells each with a constant density (though maybe varying from one shell to another) then the total force at the external point Q is the sum of the forces due to each shell and equals that due to the whole mass of the sphere gathered into a point at its own centre.

The result for a single shell is impressive enough in its simplicity; what is truly remarkable is that one can superpose the contributions of all the component shells – that the material buried thousands of miles deep in a planet is able to exert its influence on the attraction at an external point as if there were no intervening matter – yet all this is true as accurately as can be verified by experiment.

On this same theme of superposition, another example which we take for granted is that of electromagnetic radiations such as light rays in space. A ray of light from a star may have spent thousands of years on its way; at every point of its immense journey it has to intersect rays from all the other stars pursuing their own individual tracks, and this happens not once but continually. Yet this ray of starlight, itself a complex aggregate of different frequencies and amplitudes, arrives intact and for the most part just as if it had encountered nothing on its way. And in the depths of space this endless criss-crossing traffic of light rays makes little impression on the cold silence.

Here is a group of rather strange results which belong to that ideal world of elementary mechanics with its peculiar jargon.

Appendix 1. The attraction due to a sphere

1. Suppose that somewhere there is a cold homogeneous non-rotating solid spherical planet devoid of any atmosphere, having the same radius as the earth and a uniform density ρ equal to the mean density of the earth. The radius being r, its mass M would be $\frac{4}{3}\pi\rho r^3$. The force on a unit mass at its surface would by Theorem 2 be

$$G.\frac{M}{r^2} = \frac{4}{3}\pi\rho G\frac{r^3}{r^2} = \frac{4}{3}\pi\rho Gr = g, \text{ the surface gravity.}$$

Incidentally this proves that the surface gravity is proportional to planet radius, the density being uniform and constant.

2. If the unit mass were placed in a small cavity within the sphere and at a distance x from the centre, it would be inside some component shells which would exert no force on it by Theorem 1, but outside shells having a radius less than x, so the net force on it would be $G.\frac{4}{3}\pi\rho x^3/x^2$

$$= \frac{4}{3}\pi\rho Gx = \frac{g}{r}x \text{ towards the centre.}$$

Therefore if a tunnel were drilled through a diameter of this sphere, the motion equation of a mass m at distance x from the centre would be:

$$\text{force} = \text{mass} \times \text{acceleration}; \quad -m\frac{g}{r}x = m\frac{d^2x}{dt^2}; \quad \frac{d^2x}{dt^2} = -\frac{g}{r}x$$

which represents simple harmonic motion with period $2\pi\sqrt{(r/g)}$.

3. A skimming satellite circling this sphere would have a velocity v given by $v^2/r = g$. If t is the orbit time, $v = 2\pi r/t$ and so

$$v^2 = rg = \frac{4\pi^2 r^2}{t^2}; \quad t = 2\pi\sqrt{\frac{r}{g}},$$

the same time as for the tunnel vehicle.

4. So if the satellite flew over the tunnel mouth at the moment the vehicle was dropped down the tunnel, the two would again be together at the far side. Passengers travelling either by tunnel or satellite would be weightless throughout the entire journey; the tunnel vehicle would reach a greatest speed of 8 km per second on

123

passing the centre of the sphere, this being the constant speed of the satellite, and the time to cross the sphere by either method would be 42·5 minutes. The simple harmonic motion of the vehicle is just the projection on the diameter of the sphere of the uniform circular motion of the satellite.

5. It can also be verified that if any two points on the surface of the sphere were connected by a straight tunnel which would be a chord of the sphere, and equipped with a frictionless railway, then a carriage would without any need for an engine again execute simple harmonic motion between the ends of the tunnel, and all such journeys regardless of tunnel length would again take 42·5 minutes which would make timetabling particularly easy.

6. Lastly, the timetabling remains the same for spheres of other sizes provided they have the same uniform density as before. This follows either from the concealment principle or from the fact that r/g is independent of length if the density is constant. So the diametral tunnel effect could be demonstrated using quite a small earth-dense sphere, say 1 metre in diameter and with a hole drilled through its centre, and placed without residual rotation relative to the stars in some quiet corner of space. A ball bearing carefully released from rest at one end of the hole would appear at the other end after 42·5 minutes, only to disappear slowly on its return crossing. At least out there one would never be without the time!

Appendix 2. Spherical trigonometry
✳ ✳ ✳

The spherical triangle. Given three points A, B, C on a sphere centre O and with a radius of 1 unit, let each pair of points be joined by the shorter of the two great circle arcs which pass through them; the result is a *spherical triangle* ABC (Figure A2.1).

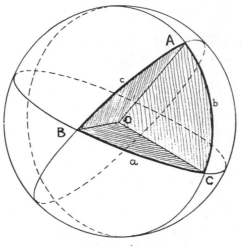

Figure A2.1

The *sides* BC, CA, AB of the spherical triangle are proportional to the central angles BOC, COA, AOB and are denoted by a, b, c.

The *angles* of the spherical triangle are those between the pairs of planes AOB, AOC; BOC, BOA; COA, COB and are denoted by A, B, C.

The cosine formula. Choose coordinate axes x, y, z as in Figure A2.2, that is, with Oz coinciding with OA and the plane zOx

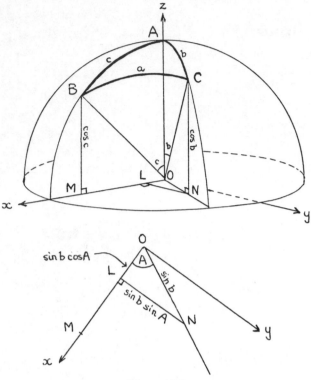

Figure A2.2

coinciding with the plane AOB. It will be seen from the figure that the vertices of the spherical triangle ABC have coordinates A(0, 0, 1), B(sin c, 0, cos c) C(sin b cos A, sin b sin A, cos b). The vectors

$$\vec{OB}= \begin{pmatrix} \sin c \\ 0 \\ \cos c \end{pmatrix} \quad \text{and} \quad \vec{OC}= \begin{pmatrix} \sin b \cos A \\ \sin b \sin A \\ \cos b \end{pmatrix}$$

have a scalar product which may be written in two ways:

$$\vec{OB} \cdot \vec{OC}=1\times1\times\cos a=(\sin c \sin b \cos A)+(0)+(\cos c \cos b).$$

Therefore cos a=cos b cos c+sin b sin c cos A.

126

Appendix 2. Spherical trigonometry

The sine formula. The perpendicular distance from C to plane AOB is equal and parallel to NL.

The volume of plane tetrahedron ABCO is

$$\tfrac{1}{3}\times(\text{base area})\times(\text{altitude})=\tfrac{1}{3}\times(\text{plane triangle area AOB})\times\text{NL}$$

$$\tfrac{1}{3}\times\tfrac{1}{2}\times1\times1\times\sin c\times(\sin b\,\sin A)=\tfrac{1}{6}\sin b\,\sin c\,\sin A.$$

By cyclic permutation of letters it is also $\tfrac{1}{6}\sin c\,\sin a\,\sin B$ or $\tfrac{1}{6}\sin a\,\sin b\,\sin C$.

Therefore $\dfrac{\sin A}{\sin a}=\dfrac{\sin B}{\sin b}=\dfrac{\sin C}{\sin c}$

To illustrate the use of the cosine and sine rules here is a pair of examples, one terrestrial and one celestial, which make use of the same calculation details.

Find the shortest distance along a great circle between New York (40° 45′ N, 74° 0′ W) and Guam Island (13° 27′ N, 144° 45′ E) in the western Pacific Ocean, also the initial course to steer from New York, the initial course to steer from Guam on the return, and the most northerly latitude reached on the way, assuming any practical and political difficulties to have been overcome.

In Figure A2.3 let P, N, G be the north pole, New York and Guam and let Q be the most northerly point en route, so angle PQN=90°.

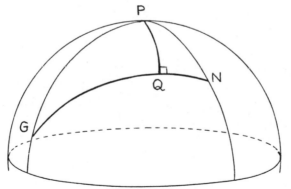

Figure A2.3

In spherical triangle PNG,
PN=90°−40° 45′=49° 15′=2955′=2955 nautical miles,
PG=90°−13° 27′=76° 33′=4593′=4593 nautical miles,
Angle GPN=360°−74° 0′−144° 45′=141° 15′=change in longitude.

To find GN

cos GN=cos 76° 33′ cos 49° 15′
$$\qquad\qquad\qquad\qquad +\sin 76° 33′ \sin 49° 15′ \cos 141° 15′$$
GN=115° 01′=6901′;   the distance is 6900 nautical miles.

To find angle GNP

$$\frac{\sin N}{\sin 76° 33′} = \frac{\sin 141° 15′}{\sin 115° 01′}; \quad N=42° 12′.$$

The initial course from New York is 360°−42° 12′=317° 48′, roughly north-west.

To find angle NGP

$$\frac{\sin G}{\sin 49° 15′} = \frac{\sin 141° 15′}{\sin 115° 01′}; \quad G=31° 33′.$$

The initial course for the return is 031° 33′.

To find the most northerly point
In spherical triangle PQN,

$$\frac{\sin PQ}{\sin 42°12′} = \frac{\sin 49° 15′}{\sin 90°}; \quad PQ=30° 35′.$$

The latitude of Q is 90°−30° 35′=59° 25′ N.

The related celestial example concerns the sun–moon paradox mentioned in Chapter 9. Suppose the sun is due west, azimuth 270°, at the low altitude of 13° 27′, and at the same time the moon is roughly south-east, azimuth 128° 45′, at an altitude of 40° 45′. The difference in azimuths being 141° 15′, the situation is again as in Figure A2.3 where now P is the zenith, G is the sun and N is the moon.

From the previous working, sunlight appears to strike the moon along a direction at 42° 12′ to the observer's vertical, having left the sun along a direction at 31° 33′ to the vertical. Also the

128

Appendix 2. Spherical trigonometry

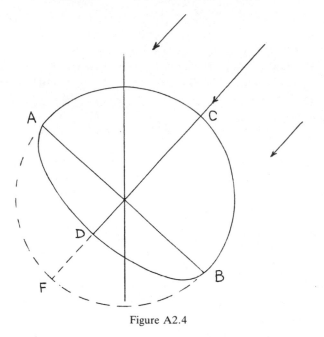

Figure A2.4

elongation of the moon is GN=115° and its phase or fraction illuminated is (1−cos 115°)/2=0·71. The moon's appearance in relation to the local vertical can now be sketched as in Figure A2.4. If the diameter CF is taken as one unit then CD=0.71 and a semi-ellipse ADB is fitted on to the semi-circle ACB.

The following problem is worked in the same way as the first example. Find the great circle distance in nautical miles between Prestwick (55° 30′ N, 4° 38′ W) and Vancouver (49° 20′ N, 123° 10′ W) and the initial course to steer from Prestwick, also the initial course from Vancouver on the return and the most northerly latitude reached on the way.

(3799 nautical miles; 325° 39′; 028° 37′; 70° 32′)

This final terrestrial example is rather different. Show that there are two points on the earth that are equidistant along great circle arcs from Edinburgh (56° N, 3° W), Hawaii (20° N, 155° W)

129

and the south pole (90° S, −); one is in a remote part of South America (5° 52′ S, 73° 25′ W) and 5048 nautical miles from each, whilst the other is in the South China Sea (5° 52′ N, 106° 35′ E) and 5752 nautical miles from each. (Use the two 'isosceles' spherical triangles each of which involves the south pole and the required point ($x°$ N, $y°$ W).)

Some celestial uses of the cosine formula

Figure A2.5 shows the celestial sphere which rotates in the sense of the arrows from the point of view of a northern observer at O. The cardinal points are N, E, S, W and Z is the zenith. P

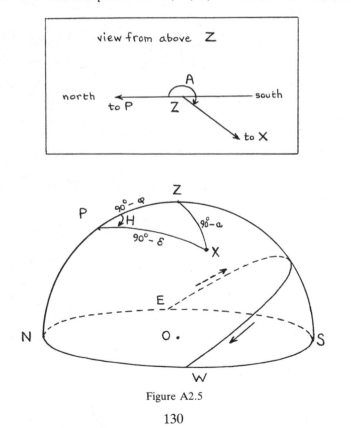

Figure A2.5

Appendix 2. Spherical trigonometry

is the celestial pole, X is an object in general position and PZX is a spherical triangle.

If the geographical latitude is ϕ then the zenith distance PZ of the pole is $(90°-\phi)$. If the object has declination δ then the polar distance PX is $(90°-\delta)$ and if the altitude or angle of elevation of the object above the horizon is a then the zenith distance of X is XZ $(90°-a)$.

The angle ZPX labelled H on the diagram is proportional to the time elapsed since the upper passage of X across the observer's meridian PZS. It is called the *hour angle* of X and often measured in sidereal hours at the usual rate of $15°=1^h$. Hour angle increases steadily from 0^h to 24^h or from $0°$ to $360°$ between successive meridian passages of X.

Lastly, angle A in the inset figure is the *azimuth* of X, being measured from north in the sense NESWN from $0°$ to $360°$. Just as δ and H are a pair of coordinates for X referred to the pole and equator, so a and A are a pair referred to the zenith and horizon. From the inset diagram it will be seen that angle PZX of the spherical triangle on the main diagram is $(360°-A)$. We are now ready to apply the cosine formula to this triangle. First,

$$\cos (90°-a)=\cos (90°-\delta) \cos (90°-\phi)$$
$$+\sin (90°-\delta) \sin (90°-\phi) \cos H$$

which simplifies to

$$\sin a=\sin \delta \sin \phi+\cos \delta \cos \phi \cos H \qquad (A)$$

Again,

$$\cos (90°-\delta)=\cos (90°-a) \cos (90°-\phi)$$
$$+\sin (90°-a) \sin (90°-\phi) \cos (360°-A)$$

which simplifies to

$$\sin \delta=\sin a \sin \phi+\cos a \cos \phi \cos A \qquad (B)$$

Example 1. Estimate the maximum duration of sunshine on the longest day in latitude $40°$, ignoring atmospheric refraction.

For the sun at noon, $H=0$. For the setting sun, $a=0$ and H is required. In result (A) write $\phi=40°$ and $\delta=23·5°$.

Do-it-yourself Astronomy

$$0 = \sin 23{\cdot}5° \sin 40° + \cos 23{\cdot}5° \cos 40° \cos H$$

$$\cos H = -\frac{\sin 23{\cdot}5° \sin 40°}{\cos 23{\cdot}5° \cos 40°}; \quad H = 111° = 7^h\ 26^m.$$

This being the sidereal time from noon to sunset, the maximum period of sunshine on that day is about $14^h\ 50^m$.

Example 2. (relevant to Chapter 8) Estimate the altitude of the sun when due west on 10 August in latitude 56° N.

10 August is 142 days after 21 March, so the sun's approximate declination would be $23{\cdot}5 \times \sin (142 \times 360/365)$ degrees $= 15{\cdot}1°$.

In result (B) write $\phi = 56°$ and $A = 270°$.

$$\sin 15{\cdot}1° = \sin a \sin 56° + \cos a \cos 56° \cos 270°.$$

The calculation happens to be easy because $\cos 270°$ is zero.

$$\sin a = \frac{\sin 15{\cdot}1°}{\sin 56°}; \quad a = 18{\cdot}3°.$$

Example 3. (also relevant to Chapter 8) To find the moment when the stars Rigel and Bellatrix in Orion are in vertical alignment seen from latitude 56° N, as in Figure 8.3.

The coordinates of these stars are:

Rigel   Right ascension $5^h\ 14^m$,   Declination 8° 13′ S

Bellatrix Right ascension $5^h\ 24{\cdot}5^m$,   Declination 6° 20′ N

The difference in right ascension is $10{\cdot}5^m$ or $2{\cdot}63°$; this is also the difference in hour angle at any instant. Let these hour angles in degrees be H for Bellatrix and $(H+2{\cdot}63)$ for Rigel which crosses the meridian ahead of Bellatrix. The problem is to find the value of H which makes the azimuths A equal and it lends itself to a step by step solution with a personal computer.

First choose a starting value for H, say 12, and then calculate a and hence A for Bellatrix:

$$a = \arcsin\left[\sin 6{\cdot}33° \sin 56° + \cos 6{\cdot}33° \cos 56° \cos H\right] \quad \text{(using (A))}$$

$$A = 360° - \arccos\left[\frac{\sin 6{\cdot}33° - \sin a \sin 56°}{\cos a \cos 56°}\right] \quad \text{(using (B))}$$

132

Appendix 2. Spherical trigonometry

The reason for subtraction from 360° is that the expected azimuth is between south and west, that is between 180° and 270°, so that cos A is negative. But a calculator or computer on meeting a negative cosine will return the numerically smaller solution between 90° and 180°.

Likewise calculate a and hence A for Rigel:

$$a = \arcsin [\sin (-8\cdot 22°) \sin 56°$$
$$+ \cos (-8\cdot 22°) \cos 56° \cos (H + 2\cdot 63)]$$

$$A = 360° - \arccos \left[\frac{\sin (-8\cdot 22°) - \sin a \sin 56°}{\cos a \cos 56°} \right].$$

Record these azimuths, select the next value of H, and repeat the process.

Here are some results which show how Bellatrix slowly wins the race.

H	A (Bellatrix)	A (Rigel)
12	195·52	195·96
13	196·79	197·03
14	198·05	198·11
15	199·31	199·18

Example 4. The occultation of Mercury by Venus, May 1737. Some further details based on the values given in Note 7.4.

(a) To find the time of sunset. In the formula

$$\sin a = \sin \delta \sin \phi + \cos \delta \cos \phi \cos H$$

write $a = 0$; $\phi = 51\cdot 5°$; $\delta = 22\cdot 6°$. Then $H = 121\cdot 55° = 8^h\ 6^m$ (using $1^h = 15°$). Sunset was at $20^h\ 06^m$ local solar time, ignoring refraction.

(b) To find the hour angle (HA) of the planets at the third observation.

RA Planets	$5^h\ 58^m$	HA Sun	$9^h\ 51^m$ (local solar time p.m.)
RA Sun	$4^h\ 23^m$	subtract	$1^h\ 35^m$
Difference	$1^h\ 35^m$	HA Planets	$8^h\ 16^m = 124°$ (using $1^h = 15°$)

(c) To find the altitude of the planets at the third observation. In the formula sin a=sin δ sin ϕ+cos δ cos ϕ cos H write

$$\delta=25\cdot6°; \quad \phi=51\cdot5°; \quad H=124°.$$

sin a=sin 25·6° sin 51·5°+cos 25·6° cos 51·5° cos 124°; a=1·4°.

In fact atmospheric refraction would increase this altitude to about 1·7°.

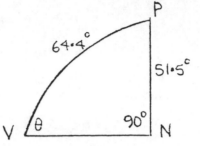

Figure A2.6

(d) To find the angle of descent of the planets when setting. In this spherical triangle, P is the north celestial pole, N is the north horizon point and V is Venus. Then arc PN=51·5°; arc PV=90°−25·6°=64·4°. By the sine rule,

$$\frac{\sin\theta}{\sin 51\cdot5°} = \frac{\sin 90°}{\sin 64\cdot4°}; \quad \theta=60\cdot2°$$

so the required angle is approximately 30°.

(e) To find the azimuth of the planets when setting. In the formula sin δ=sin a sin ϕ+cos a cos ϕ cos A write a=1·4°; ϕ=51·5°; δ=25·6°. Then cos A=0·6636. Selecting the north-west region, A=360°−48·4°=312° approximately.

References

✳ ✳ ✳

Many of the positions of the sun and planets were derived from the program in BASIC published in book form:
P. Duffett-Smith: *Astronomy with your Personal Computer*
Cambridge University Press, 1985

✳

The numbered references in the text are:
1
C. Hutton, G. Shaw and R. Pearson
Philosophical Transactions of the Royal Society, London
abridged, vol. 8, p. 590: London, 1809
2
Nature, vol. 8, p. 117, June 1873
3
Philosophical Transactions, abridged, vol. 3, p. 448
4
Philosophical Transactions, abridged, vol. 6, p. 243
5
D. Gill: *Monthly Notices of the Royal Astronomical Society* (1877)
6
W. W. Rouse Ball: *A Short Account of the History of Mathematics*
(Dover reprint), p. 332
7
W. W. Rouse Ball: *op. cit.*, p. 333
8
A. R. Hall and M. B. Hall
Unpublished Papers of Sir Isaac Newton: A Selection from the Portsmouth Collection
Cambridge University Press, 1962, p. 381

Index

✳ ✳ ✳

Index